# Natural Gas Transmission and Distribution Business

# Natural Gas Transmission and Distribution Business

Pramod Paliwal
Sudhir Yadav

CRC Press
Taylor & Francis Group
Boca Raton London New York

CRC Press is an imprint of the
Taylor & Francis Group, an **informa** business

CRC Press
Taylor & Francis Group
6000 Broken Sound Parkway NW, Suite 300
Boca Raton, FL 33487-2742

First issued in paperback 2020



ISBN-13: 978-1-138-59830-0 (hbk)
ISBN-13: 978-0-367-65658-4 (pbk)



**Library of Congress Cataloging-in-Publication Data**

Names: Paliwal, Pramod, author. | Yadav, Sudhir, author.
Title: Natural gas transmission and distribution business / Pramod Paliwal and Sudhir Yadav.
Description: Boca Raton, FL : CRC Press, 2018.
Identifiers: LCCN 2018046026 | ISBN 9781138598300 (hardback: alk. paper)
Subjects: LCSH: Gas industry. | Gas companies. | Natural gas pipelines.
Classification: LCC HD9581.A2 P36 2018 | DDC 363.6/3--dc23
LC record available at https://lccn.loc.gov/2018046026

**Visit the Taylor & Francis Web site at**
**http://www.taylorandfrancis.com**

**and the CRC Press Web site at**
**http://www.crcpress.com**

*To my parents, wife, and two wonderful children.*

**—Pramod Paliwal**

*To my parents and family.*

**—Sudhir Yadav**

# Contents

# Foreword

Natural gas has been an energy source for well over a hundred years and globally most countries now use natural gas. In over 30 years of experience working in the gas industry starting with British Gas as an apprentice, I have seen significant change from working in different regions of the world on projects. The British Gas heritage is greatly admired and also the knowledge and research which British Gas pioneered has been invaluable to today's gas systems. I often come across British Gas procedures and practices in my travels.

I have had the great honor of chairing many gas industry conferences and more recently serving in the prestigious role as President of the Institution of Gas Engineers and Managers (IGEM) in the United Kingdom. These roles provided me with some exceptional experiences and privileges within the gas industry. One of the key points I observed is the passion for learning and knowledge development within the industry, which plays a pivotal role in research and further expansion.

It is vital that we capture improvements and advancements within the gas industry to enable others to continue learning. Continuing improvement and education are essential for all levels, from graduates to experienced engineers working in the energy arena.

I had the special honor of knowing Dr Pramod Paliwal for several years through our career paths in the gas industry and through sharing ideas and knowledge. He and his coauthor have undertaken a momentous amount of time reaching the level of understanding that they have of the gas value chain. They are now recognized as thought leaders in the field, which is well deserved and an invaluable source of reference, a true legacy for future testimonials.

This book covers the value chain of natural gas from transmission to the end user. In today's current climate the landscape of the regulatory framework is important and, with the varying maturity of markets, illustrates the differing structures in place to meet the regulatory goals.

This book examines the business areas of transmission and distribution and the authors aligned the chapters to the differing operating factors from operation to maintenance areas. They drew upon a significant amount of information from various countries and jurisdictions to provide the key areas and objectives for the subject areas.

The immense changes in research and development in the gas industry have been vast and the challenges of technological advancements are great and, therefore, critical to the business areas.

Finally, I would like to give a few words of encouragement to the current generation of energy and gas industry professionals and their successors:

students and scholars. Learning is a never-ending process and the gas industry will continue to develop for many years, adapting to new and varying challenges. It will be gas engineers and managers who will drive its pathway.

**Andy Cummings**
*DNV GL—Oil & Gas*
*Institution of Gas Engineers and Managers (IGEM), United Kingdom*

***Andy Cummings,*** *BEng (Hons) CEng FIGEM MIET. A chartered engineer with over 30 years of experience in the gas industry. Recently, he undertook a very prestigious and exciting role as President of the Institution of Gas Engineers and Managers in the United Kingdom for 1 year. He considered it an honor and a pleasure to engage in the industry with this role, and it provided a great insight into some of the challenges facing the energy industry for the future. He has been involved in a number of city gas projects from conceptual and feasibility studies and the development of the city gas distribution technical capability and knowledge within the business. His client's interfaces have included BG, PTT Natural Gas Division Thailand, Qatar Petroleum, Oman Gas Company Oman, Adnoc Gas Distribution UAE, Kuwait Oil Company, National Grid, PetroVietnam, and Energy Regulators for the development gas safety regimes. Dr Cummings is a fellow of the Institution of Gas Engineers and Managers.*

# Preface

Natural gas value chain, apart from exploration and production, also has two major components: transportation and distribution. With the vital role of ensuring the last mile of delivery of natural gas, natural gas transmission and distribution have been a growing area of interest to all concerned: industry, governments, energy policy makers, and academia. The effort of governments and policy experts on promoting the natural gas sector increases the need to have a well-documented book that deals with the business issues of natural gas transportation and distribution. So far, magazine articles, a few discussion papers, and reports by consulting companies document the major discussion on the natural gas transmission and distribution business. Thus, a book is needed to addresses these vital aspects of the natural gas business. It is with pleasure that, with Taylor & Francis taking the lead, we are taking this opportunity to write a structured book that deals with the commercial, managerial, and regulatory aspects of natural gas transportation and distribution business with an applied technical perspective wherever needed. These perspectives include reflections on project management; regulations; marketing; health, safety, and environment (HSE); and operations and maintenance (O&M). With well-structured chapters that contain, appropriate tables, graphics, and illustrations, we present a comprehensive view on this business. We have been able to do this despite the constraints of book-size because of the planning and execution of this title. This book also contains a case that gives a perspective of emerging natural gas distribution business models and how the conventional natural gas transmission and distribution business tackles these issues.

Apart from contextual material in the chapters, we have also provided sources of information to provide readers with additional readings from databases reports and so on that would benefit those who would like to delve deeper into specific issues. This book, we feel, will help all stakeholders in the oil and gas industry including professionals, business executives, techno-managerial personnel, students, and faculty members of energy sector managerial and technology programs.

**Dr. Pramod Paliwal**
**Dr. Sudhir Yadav**

# *Acknowledgments*

Throughout the development of this volume, several extraordinary individuals provided their time and support to keep us motivated, share valuable inputs, and help organize field visits. Especially, we thank the following individuals and organizations:

C.G. Gopalkrishnan

Nayan Pandya

D.M. Pestonjee

Gyanendra Kumar Sharma

Varun Patel

Taylor & Francis India

Mahanagar Gas Limited

LNG Express Private Limited

Cryogas Industry Group

Former & Current students of School of Petroleum Management, Pandit Deendayal Petroleum University

We cannot thank the editorial, publication, and marketing team at Taylor & Francis Books India Pvt. Ltd. enough. With a deep sense of appreciation, we thank Dr. Gagandeep Singh and Mouli Sharma.

Furthermore, Dr. Pramod Paliwal and Dr. Sudhir Yadav are extremely thankful to one and all at the School of Petroleum Management, Pandit Deendayal Petroleum University, and Gandhinagar Gujarat for providing the right academic environment and motivation to work on this ambitious project.

Dr. Pramod Paliwal thanks his wife Shally, daughter Gargie, and son Neil for bearing with him throughout the process of writing this volume and for keeping his spirits high.

Dr. Sudhir Yadav acknowledges the constant support of his wife Sapna and sons, Siddharth and Karan, during this book project.

# Authors

**Pramod Paliwal** has over 27 years of corporate and academic experience. With an MBA and PhD in management, he is a Fellow of The Chartered Institute of Marketing, UK. He has been with the School of Petroleum Management, Pandit Deendayal Petroleum University, Gandhinagar-Gujarat, India, since 2006. In the profession since 1991, Dr. Paliwal's earlier associations were with the ACC Limited (a subsidiary of Holcim, Switzerland), Pacific Institute of Management, M.L.S. University, India, and the India Offshore Campus of Oxford Brookes University, UK. He has been featured in the 30th Pearl Anniversary Edition of Marquis *Who's Who in the World* 2013. In 2017, he was selected to serve on the panel of Chair of Indian Studies Abroad by Indian Council for Cultural Relations (ICCR), MEA, Government of India. He is associated with the international crowdsourcing consulting firm Wikistrat as a senior analyst in the area of energy. He also speaks and writes on sustainable development and public policy.

Given his academic, professional, and research interest in sustainability, sustainable marketing, and energy and resources sector, he presented, has been a panelist, and conducted workshops at international events in India, Europe, the Middle East, North Africa, Southeast Asia, and the United States. He has published on sustainability, marketing, and energy sector issues in refereed international journals and also a book published by McGraw-Hill.

**Sudhir Yadav** has more than 28 years of experience including about seven years in industry and 20 years in academia. He has been with the School of Petroleum Management, Pandit Deendayal Petroleum University, Gandhinagar-Gujarat, India, since 2006. As an industry professional, he worked in the marketing and international business divisions of various companies. Dr. Yadav developed and delivered academic and training modules in the areas of oil and gas operations management, oil and gas project management, and the oil- and gas-value chain. His case studies and publications in oil and gas management have been acclaimed by learners. He conducted in-company and executive management programs for corporate executives. In the past, he contributed to the Natural Gas Technical Skill Development initiatives of Petroleum and Natural Gas Regulatory Board (PNGRB). He visited the United States, Europe, the Middle East, and Southeast Asia for his academic and professional assignments. Dr. Yadav delivered oil and gas training programs in the Middle East. Dr. Yadav also published a book with McGraw-Hill.

# 1

# *Natural Gas Industry: An Introduction*

Energy is vital to all spheres of human activity. In this modern world, where technology plays such an important role in our lives, the importance of energy has never been so keenly felt. The world's primary energy comes from various sources of which natural gas is a vital component. It is one of the cleanest, safest, and most versatile of all energy sources.

This chapter will discuss the rudimentary aspects of natural gas as a commodity; its significance among other fuels; and how natural gas is produced, processed, and transported to distribution companies. In addition, the chapter will cover the typicality and challenges of natural gas transmission and distribution along with preliminary aspects of safety.

## 1.1 Natural Gas: An Introduction

Natural gas, a hydrocarbon, is colorless, shapeless, and odorless in its nascent form. Natural gas is a combustible molecule found in various regions of the world (both onshore and offshore) and is a clean fuel relative to other sources of primary energy (Figure 1.1).

Natural gas, though, is not always used as a fuel (as it has other chemical uses also). Still it is primarily considered as a hydrocarbon fuel. It is this need for energy that has elevated natural gas to such a level of importance in our society and in our lives.

Natural gas is a combustible mixture of various hydrocarbon gases. Although natural gas is formed primarily of methane, it can also include other components such as ethane, propane, butane, and pentane. Depending upon sources, the composition of natural gas can differ. However, the following graphic gives an idea about the typical composition of natural gas in its raw form (Figure 1.2).

Methane is a molecule composed of one carbon atom and four hydrogen atoms and is referred to as $CH_4$. The distinctive pungent smell associated with natural gas is due to an odorant called Mercaptan that is added to the gas before it is delivered to the end-user. Mercaptan acts as a leak-detector.

Ethane, propane, and the other hydrocarbons usually associated with natural gas have somewhat different chemical formulas. Natural gas bereft of

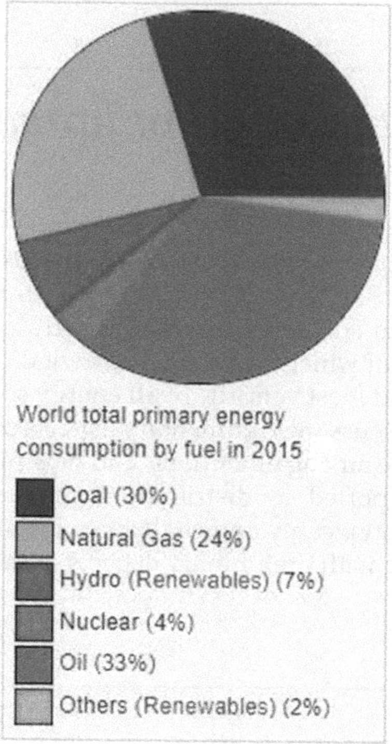


**FIGURE 1.1**
World energy basket and natural gas. (From https://en.wikipedia.org/wiki/World_energy_consumption.)

| | |
|---|---|
| Methane | 70-90% |
| Ethane/Propane/Butane | 0-20% |
| Carbon Dioxide | 0-8% |
| Oxygen | 0-0.2% |
| Nitrogen | 0-5% |
| Hydrogen Sulphide | 0-5% |
| Rare gases | trace |

**FIGURE 1.2**
Composition of natural gas.

other commonly associated hydrocarbons is generally termed as dry because it is nearly pure methane (which is very rare). In the presence of other hydrocarbons, it is termed as wet.

Natural gas has multiple uses, most commonly residential, commercial, and industrial. Natural gas is found in reservoirs beneath the ground and the seabed. If it is explored and produced as a part of crude oil drilling process, then such natural gas is termed as associated (natural) gas. However, it is also found, explored, and produced independent of crude oil reservoirs. Exploration and production (E&P) companies deploy modern techniques to locate the reserves of natural gas and use sophisticated technology to explore and produce the commodity. Natural gas once it reaches the wellhead (from below the ground) is subjected to refinement (cleaning) to eradicate impurities such as moisture, other gases, sand, and chemical compounds. Depending upon the economic viability, some hydrocarbons (such as propane and butane) may be removed and sold separately. After refining, the clean natural gas is ready to be transmitted through a network of pipelines to reach the end users. The physical measurement of natural gas can be done using different metrics. As a commodity, it can be measured by the volume (at standard, temperatures, and pressures (STP)), generally expressed in cubic feet. Production and distribution companies commonly measure natural gas in cubic feet meaning millions of cubic feet (MMcf), or trillions of cubic feet (Tcf). Daily production and consumption of natural gas can also be measured in standard cubic meters (SCM) per day (SCMD). Larger volumes are expressed in million standard cubic meters per day (MMSCMD). MM stands for Latin *mille mille* (thousand into thousand) meaning million. Natural gas can also be measured by potential energy output (i.e., calorific value). As in the case of other forms of energy, natural gas also is commonly measured and expressed in British thermal units (Btu). One Btu is the amount of natural gas that will produce enough energy to heat 1 pound of water by 1 degree at normal pressure (Figure 1.3).

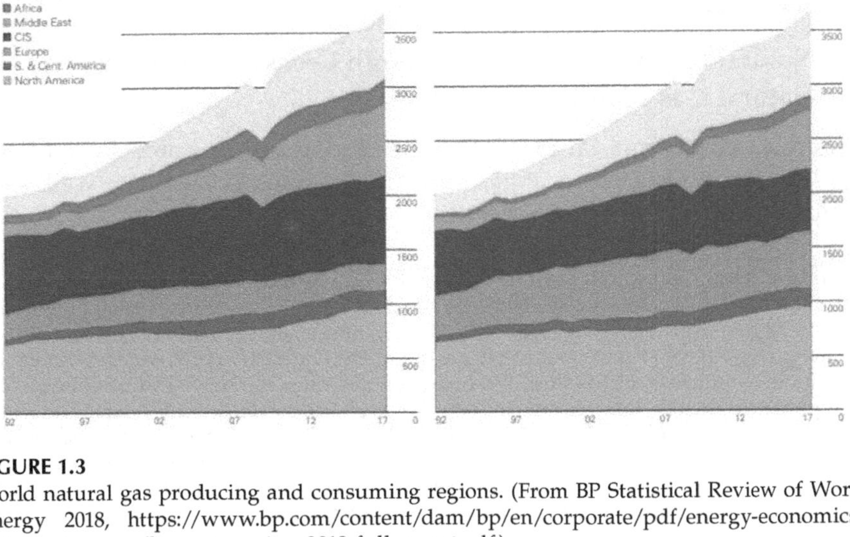


**FIGURE 1.3**
World natural gas producing and consuming regions. (From BP Statistical Review of World Energy 2018, https://www.bp.com/content/dam/bp/en/corporate/pdf/energy-economics/statistical-review/bp-stats-review-2018-full-report.pdf.)

## 1.2 Natural Gas Exploration and Production

In contrast to the past, the last couple of decades have witnessed great changes in natural gas (and crude oil) exploration and production activities. These changes have been made possible with the availability of modern sophisticated technology and a good amount of research and development in the E&P domain. After ascertaining the commercial viability of natural gas reserves (during the initial exploration process), it is brought to surface to be processed and transported through pipelines to the requisite destinations (i.e., either variety of consumers or to Liquefaction Facilities) (for Liquefied Natural Gas-LNG[1] production).

Effective and efficient movement of natural gas from producing regions to consumption (and Liquefaction Infrastructure) regions requires a widespread and highly structured transportation system. In many instances, natural gas produced from a specific well travels vast distances to reach its point of use. The natural gas transportation system comprises a composite network of pipelines designed for smooth transportation of natural gas from the wellhead to the consumption points.

The transportation system consists of (a) a gas gathering system (GGS), (b) an interstate or cross-country pipeline system, and (c) the city gas distribution network. GGS facilitates the gathering of the natural gas produced from various producing wells nearby so that the gas can be pumped into high-pressure cross-country pipelines for further transmission at long distances. GGS also is a point where natural gas is subjected to some chemical treatment to tackle high sulfur or carbon dioxide and some other impurities that may be present in natural gas.

## 1.3 Natural Gas Transmission Through Cross-Country Natural Gas Pipelines

The cross-country natural gas pipeline network transports processed natural gas from processing plants at gas gathering stations located near wellheads to consumers located at the last mile.

1 **Liquefied natural gas (LNG)** is natural gas (predominantly methane, $CH_4$, with some mixture of ethane $C_2H_6$) that has been cooled down to liquid form for ease and safety of non-pressurized storage or transport. It takes up about 1/600th the volume of natural gas in the gaseous state (at standard conditions for temperature and pressure). The liquefaction process involves removal of certain components, such as dust, acid gases, helium, water, and heavy hydrocarbons, which could cause difficulty downstream. The natural gas is then condensed into a liquid at close to atmospheric pressure by cooling it to approximately –162°C (–260°F); maximum transport pressure is set at around 25 kPa (4 psi). Natural gas is mainly converted to LNG for transport over the seas where laying pipelines is not feasible technically and economically.
*Source:* https://en.wikipedia.org/wiki/Liquefied_natural_gas

In a way, the cross-country pipelines are a super highway for natural gas transportation. Natural Gas is transported through these pipelines at very high pressures ranging from 200 to 1,500 pounds psi. Such high pressures are important for propulsion of natural gas at high speeds through the cross-country pipeline network. Compressor technology is used to maintain requisite pressures in the pipelines.

## 1.4 Transmission Pipes

Since natural gas transportation at high speeds is a complex phenomenon, it is important that the cross-country pipelines operate with high degree of reliability and integrity. Pipeline material and associated components, parts, and other machinery have an important role to play in this operation.

Transmission pipes made generally of high-quality stress-bearing carbon steel measuring anywhere from 6 to 48 inches in diameter. These pipelines are designed and manufactured to meet the high engineering standards set by agencies like the American Petroleum Institute (API) and The American Society of Mechanical Engineers (ASME).

Although the main transmission pipes in the cross-country transportation network are usually between 16 and 48 inches in diameter, the lateral pipelines, which deliver natural gas to or from the mainline, are usually between 6 and 16 inches in diameter. Most major cross-country pipelines range from 24 to 36 inches in diameter. To put it in perspective, the distribution network pipelines are made of special quality plastic due to the need for flexibility, versatility, and ease of maintenance.

As also referred previously, the cross-country transmission pipelines are manufactured using special quality carbon steel material and sophisticated technology. This construction is important to ensure that these pipelines work with highest degree of integrity to meet the requirements of pressure handling and strength of the pipeline. Moreover, since the commodity (i.e., natural gas) being transported is of such physical and chemical characteristics, adherence to advanced quality standards during manufacturing is extremely crucial.

Specialized coating techniques are used to prevent corrosion in the pipelines from outside and inside. Apart from coating techniques like epoxy coating, cathodic protection also is commonly used to prevent the pipelines from corrosion. Cathodic protection is the technique used to protect the steel pipelines wherein, using the principles of electricity, the pipeline metal is made to act as an electric cathode to prevent corrosion under the influence of moisture.

## 1.5 Compressor Stations

To facilitate the transmission of natural gas at high pressures in the cross-country pipelines, adequate pipeline pressure must be consistently managed in the pipelines. A series of compressor stations usually placed at 60–160 kilometer intervals along the pipeline manage pipeline pressure. Natural gas is compressed at these compressor stations by using machines (such as engines, turbines, and motors) and technology generally used at such compressor stations.

Technologies are available wherein the turbine compressors draw their energy by utilizing some quantity of the natural gas that these turbines compress. The turbine in this case serves to operate a centrifugal compressor, which consists of a fan that compresses and pumps the natural gas through the pipeline. On the other hand, some compressor stations have arrangements of electric motor to turn the same type of centrifugal compressor. Although such compression does not require the use of any of the natural gas from the pipe, it does require a reliable source of electricity along the pipeline.

In some cases, compressor stations are powered by natural gas-based engines. These engines, which compress the natural gas, also simultaneously draw their natural gas from the pipeline.

Compressor stations in addition to compressing pipeline natural gas also sometimes function as points to trap and separate (by using scrubbers and filters) any kind of residual moisture content and superfluous particles from the natural gas. This process is in addition to the function primarily carried out at gas gathering stations.

## 1.6 Metering Stations

Measurement of the flowing natural gas is an extremely important task. A well-designed metering system can carry out this task and hence, simultaneous to the transportation, metering stations accomplish this task. Measurement process through these metering stations with the help of specialized online meters is done in such way that does not disrupt the flow of natural gas. Metering stations are necessary from the control point of view because this metering also has commercial implications.

## 1.7 Valves

Like any other well-established pipeline system, a valve sub-system plays a very important role in the cross-country pipeline system for natural gas transmission. Valves are used to regulate the flow of gas in the pipeline.

Need for regulating the flow arises due to operation and maintenance (O&M) reasons and thus valves, which are placed at various intervals of the pipeline, are an essential mechanism in the system. All aspects related to valves, from the design to the choice of material and physical production, must adhere to relevant manufacturing and safety codes.

## 1.8 Monitoring Stations and Supervisory Control & Data Acquisition Systems

As discussed, natural gas flowing in the cross-country transmission pipelines at high pressure and speeds is a relatively complex process. Apart from the flow of commodity, many other aspects such as safety, leak detection, managing risk and hazards, ensuring that quantities are not lost while flow, and so on are to be taken care of. Similarly, protecting the interests of both entities (i.e., the producers/merchants of natural gas and the customers at the last mile) is also very important. A well-established monitoring system thus becomes very important to ensure this protection. The monitoring system also must work with minimum human interface. That is, because given the vast lengths (running into hundreds of kilometers), it is not always possible for on-site personnel to monitor and control the system.

To carry out this task of monitoring and controlling the natural gas that is traveling through the pipeline, sophisticated centralized gas control stations are established that collect, collate, and manage data received from monitoring and compressor stations all along the pipeline. Data management, hence, has an important role in monitoring and controlling the natural gas.

Supervisory control and data acquisition (SCADA) systems accomplish this task of managing data. SCADA is a proven sophisticated communication mechanism that measures and collects data online while the gas is in flow (metering or compressor stations and valves are the origin of such data) and transmits it to the central control station. Relevant data such as pipeline flow rate, operational status, pressure, and temperature help in estimating the status of the pipeline and the commodity at any point in time. SCADA has the uniqueness of operating in real time, thus there is almost zero lag time between the measurements taken along the pipeline and data transmission to the centralized control station. This system ensures immediate corrective actions (if required).

Such corrective actions usually pertain to equipment malfunction, leaks, or any other abnormal (may be surreptitious) activity along the pipeline. As mentioned previously, some advanced SCADA systems also provide for remote diagnostics of possible faults in equipments along the pipeline, including compressor stations, allowing maintenance professionals in a centralized control station to take corrective actions.

## 1.9 Pipeline Construction

With the gradual increase in production and consumption of natural gas in most consuming regions, the cross-country pipeline network also keeps expanding. The network comparatively must grow faster because capital intensive projects have their own project management challenges and, hence, they need to not only have scalability but also must be ready whenever natural gas is ready to flow. In many markets around the world, all cross-country pipelines have some excess capacity at any given point of time, which is also because pipeline capacity creation is a time intensive project and there is less scope of any flexibility to tweak with capacities in the short to medium terms.

The construction of natural gas pipelines (both cross-country and distribution network) involves significant planning and groundwork.

Detailed feasibility analysis including the survey of potentially affected communities, natural resources, farmlands, water bodies, terrains, and so on must be carried out. In addition, a comprehensive Environment Impact Assessment (EIA) is also undertaken to identify and mitigate any kind of potential adverse environmental impacts due to construction of natural gas pipelines. Analysis must also ensure that any existing public or private infrastructure is not affected due to this activity. If there may any impact at all on farmlands and pre-existing infrastructure, then adequate compensation needs to be paid to the owners of these properties and assets.

Aspects like right of way (RoW) and right of use (RoU) also must be tackled before laying the transmission pipelines. An RoW is a strip of land that is granted, through an easement or other mechanism, for transportation purposes, such as for a path, driveway, railway track, or natural gas pipeline.

An RoU is the temporary right possessed by the owner of natural gas pipeline network to use the strip of land obtained by RoW for laying the pipelines.

Adequate care must be taken whenever a pipeline route includes bodies of water, farmlands, or forests. Pipeline construction management must ensure that there is minimum impact to these. Even this minimum impact must be made almost zero by restoring top soils, constructing underground tunnels (in case of road crossings), putting pipelines at the riverbed (with adequate coatings) in case the river falls en route, fencing, and so on (Figure 1.4).

After completing all necessary surveys, environmental studies, and obtaining necessary permissions (both private as well as statutory), the actual pipeline construction is carried out. As also indicated earlier, keeping in view the complexity of cross-country pipelines, specialized civil engineering and mechanical engineering techniques are required to undertake the pipeline construction project management. Moreover, peculiarities in the route of the

**FIGURE 1.4**
Cross-country natural gas pipeline. (From https://www.hydrocarbons-technology.com/wp-content/uploads/static-progressive/nri/hydrocarbons/Projects/Roadrunner/roadrunner-l.jpg.)

pipeline network must be tackled and care must be taken to avoid time and cost overruns in the project. Coating requirements according to needs also must be simultaneously fulfilled.

After the construction phase (including welding, coating, valves, metering system, and compression infrastructure) is completed, trial runs are carried out before the actual flow of natural gas. This trial run is done through a test called the hydrostatic test. Water is made to flow at higher pressures (equal to or greater than the pressure at which natural gas will flow in the pipeline) through the entire length of the pipeline. The hydrostatic test checks the endurance of the pipeline network along with any possibilities of leaks, breakages, fissures, and so on.

## 1.10 Pipeline Integrity Management: Inspection and Safety

To ensure the efficient and safe operation of the vast network of natural gas pipelines, O&M staff periodically inspect the pipelines for corrosion, fissures, leakages, or any other defects. This inspection is carried out by using a sophisticated technique called pigging. Pigging involves the use of smart pigs, which are intelligent robotic devices (with advanced sensor technology) that are inserted into the pipeline and made to pass through the pipelines to thoroughly check the interior surfaces of the pipe. These smart pigs can assess pipeline thickness, spherical properties, any potential indicators of corrosion, detect minute leaks due to fissures/detaching of joints, and any other fault along the interior of the pipeline that may result into obstruction in flow of gas, or create a likely safety risk in pipeline operation. Incidentally pigging is just one of the sophisticated techniques among an array of procedures and techniques that are carried out to ensure integrity of cross-country pipelines at all times.

Following is a list of the major safety precautions for natural gas pipelines:

- **Aerial and ground patrols:** Planes, drones, and ground patrolling are used to ensure that no undue construction or digging activities are taking place in proximity to the pipeline route, particularly in residential areas. Usually unauthorized construction and digging is the primary threat to pipeline safety.
- **Leak detection:** Sophisticated natural gas-detecting equipment is periodically used by pipeline maintenance teams on the surface to check for leaks. It is possible that odorants (like Mercaptan) are not always used and hence mechanical or electronic leak detection techniques are required.
- **Pipeline identifiers and markers:** Placing appropriate signage, color codes, and markings on the ground above natural gas pipelines indicate the presence of underground pipelines to the public (and other utility providers). The objective of these identifiers and markers is to mitigate the risks of (even unintentional) tampering with underground pipelines.
- **Natural gas sample collection:** O&M personnel routinely collect samples of the natural gas flowing in the pipelines to ensure its quality. If a sample is found to be contaminated, then it may be an indication of corrosion in the interior of the pipeline or the incursion of other external contaminants.
- **Preventative maintenance:** This precaution involves the routine testing of valves, metering equipment, and clearing the surface obstructions if any that may hamper smooth pipeline inspection.
- **Emergency response:** All pipeline companies have extensive emergency response and rescue teams which are always geared up for any kind of potential accidents and exigencies.

## 1.11 Storage of Natural Gas

Though not always required, natural gas, like most other commodities, can be stored for any period. The exploration, production, and transportation of natural gas is a time-consuming process, and the natural gas that reaches its destination at times may not always be needed right away. So very often it may be stored in underground storage facilities. These storage facilities can be located near consumption points that do not have a ready supply of locally produced natural gas. These underground storage facilities have their respective geological characteristics and different kinds of underground storage facilities can be used according to injecting and withdrawal requirements.

In addition, a mechanism called line pack also by default can be used to store natural gas. Line pack is a procedure for allowing more gas to enter a pipeline than is being withdrawn, thus increasing the pressure, thus packing more gas into the system and effectively creating storage. The packed gas can subsequently be withdrawn when needed. (Source: www.gasstrategies.com/industry-glossary) (Figure 1.5)

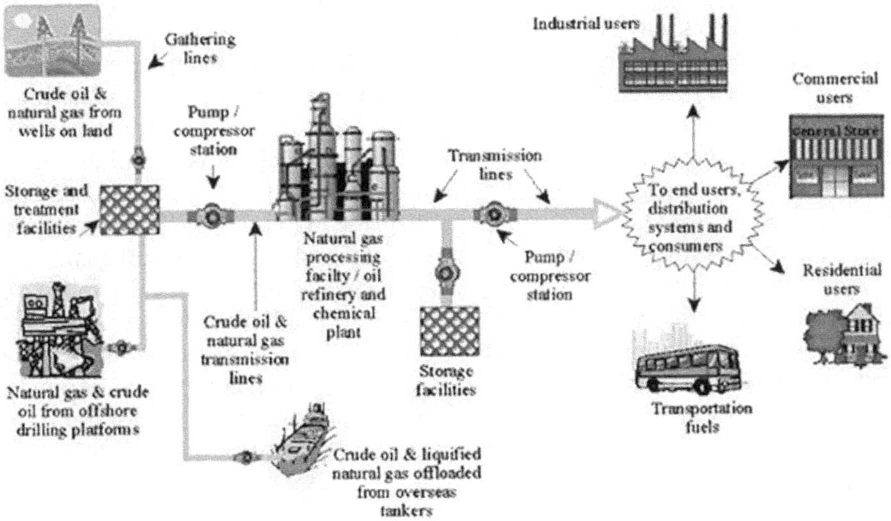


**FIGURE 1.5**
Natural gas supply system. (From Paliwal, P., *Vikalpa*, 35, 61–91, 2010.)

## 1.12 Natural Gas Distribution: An Introduction

Distribution is the last mile stage in making natural gas available to customers. Although some industrial, commercial, transportation sector (Compressed Natural Gas[2]; CNG Retail Stations), and electricity generation customers having bulk requirements at relatively high pressures may receive natural gas directly from cross-country pipelines (through a separate trunk line bifurcated from cross-country pipelines), local gas utility, also called a local distribution company (LDC) or city gas distribution (CGD) company, deliver natural gas to most other consumers. LDCs and CGD companies are utilities regulated by a concerned natural gas distribution framework, engaged in the business of delivery of natural gas to consumers within a specific geographic area (GA). These consumers usually have natural gas requirements of relatively lower quantities at very low pressures. LDCs and CGD companies are geared up to cater to these requirements using a unique CGD network that can cater to these requirements.

Local distribution companies typically distribute natural gas from delivery points located on high-pressure and bulk carriage cross-country pipelines to domestic and commercial customers through an intensive network that at times may consist of thousands of route-kilometers of small-diameter distribution pipes made of synthetic material (as compared to carbon steel that is used in manufacturing of high-pressure cross-country pipelines).

This network being specific to the consumption centers is intensive in nature and hence is also known as a reticulated network. The point of delivery where the natural gas is transferred from a cross-country transmission pipeline to the local gas distribution utility is known as the city gate station (CGS), and at times (depending upon different regulatory systems) also acts as an important market (commercial exchange) for the pricing of natural gas in large urban areas. Along with to the CGS is located a system known as district regulating station (DRS), DRS has a function to regulate (reduce) the high pressure to the requisite low pressures to suit the needs of customers of low-pressure natural gas.

Typically, utilities take possession (also known as custody in industry terminology) of the natural gas at the CGS and deliver it to each individual customer's location passing through individual consumer meters. As mentioned, this necessitates the requirement of a dense network of small-diameter distribution pipelines. Depending upon intensity of geographical areas, this network may range from a thousand to a million kilometers.

[2] **Compressed Natural Gas** (**CNG**) is a fuel which can be used in place of gasoline (petrol), diesel fuel, and propane/LPG. CNG combustion produces fewer undesirable gases than the fuels mentioned previously. CNG is made by compressing natural gas to less than 1 percent of the volume it occupies at standard atmospheric pressure. It is stored and distributed in hard containers at a pressure of 20–25 MPa (2,900–3,600 psi), usually in cylindrical or spherical shapes.
*Source:* https://en.wikipedia.org/wiki/Compressed_natural_gas

Because of the transportation infrastructure required to move natural gas to many diverse customers across a reasonably wide geographic area, distribution costs typically make up about half of natural gas costs for households and small volume customers. Although large pipelines can reduce unit costs by transmitting large volumes of natural gas, distribution companies must deliver relatively small volumes to many more different locations. Typically, transmission and distribution costs represent about half the monthly gas utility bill of a typical customer of residential natural gas, with costs of the physical natural gas commodity representing the other half of the bill.

## 1.13 Delivery of Natural Gas

The delivery of natural gas to its last mile (location of the end user) by a natural gas distribution utility (LDCs or CGD company) is like the transportation of natural gas previously discussed in the transportation section. The major difference, however, is that the distribution activity consists of carrying very small quantities of gas at much lower pressures over lesser distances to many individual customers. Also, the pipes are small diameter (compared to the pipes having large diameters in case of transmission pipelines) and are used to cater to the needs of individual customers. It would be contextual to mention here that Compressed Natural Gas (CNG)[2] Retail Stations (that cater to the transportation fuel needs of vehicular customers) are also supplied their natural gas (after compression of natural gas) from the CGS and, hence, are very much a part of the distribution network. A following chapter will provide a detailed discussion on CNG and its retailing.

Pressure regulation mechanism is carried out periodically to ascertain a smooth pipeline flow in the distribution network. Since only lesser quantities of natural gas are moved in the small diameter pipelines, the pressure needed to move the natural gas through the distribution network is substantially lower than that required in the (bulk) transmission pipelines. To put things in perspective, although natural gas traveling through cross-country pipelines may be compressed to as much as 1,500 pounds per square inch (psi), natural gas that traverses through the distribution network requires as little as 3 psi of pressurization and at the minimum only ¼ psi at the customer's meter. As mentioned previously, the natural gas for distribution is typically depressurized at or near the CGS (at the DRS), and if needed is also scrubbed and filtered (even though it has already been processed previously at the time of transmission through cross-country pipelines) to tackle moisture and unwanted particulate content, if any. In addition, Mercaptan, the odorant, is added by the LDCs or CGD companies prior to distribution. This odorant, as also mentioned earlier, is added to make the detection of leaks much easier.

Although in the earlier days, rigid steel pipe was used in CGD networks, new technology enables the use of flexible plastic and corrugated stainless steel tubing and copper tubing in place of rigid steel pipe. Using new materials in distribution pipes results in cost reduction and, in addition, also provides installation flexibility and easier operations and maintenance for both entities: utilities and consumers.

Electronic meter-reading systems having on-site meters facilitate the measurement of the consumption of natural gas by various customers in the distribution network.

Though it operates under low-pressure regimes and the quantities per GA are low, the laying down of CGD networks is not without its own share of challenges. That is because these networks almost routinely must be built in areas that have dense population. Also, a lot of coordination is required from existing utilities (electricity distribution, water supply, optical fiber network, sewage, and so on) that also have underground assets. In addition, RoW and RoU permissions are required as usual. Since most urban and semi-urban areas have jurisdiction of concerned municipal authorities, their required permissions are necessary to lay down the network. Moreover, the challenges for construction project management remain the same as the present in the case of laying down the cross-country pipelines for the transmission of natural gas, but, of course, the complexity is less.

CGD companies also use SCADA systems, comparable to those used by cross-country transmission pipeline companies. These systems can integrate gas flow control and measurement with other accounting, billing, and contract systems for a comprehensive measurement and control system for the CGD company. This system provides for precise real-time information on all aspects of the distribution network and ensures efficient and effective service to the customers in the network.

## Bibliography

Aune, F. R., R. Golombek, and S. A. C. Kittelsen. (2004). "Does increased extraction of natural gas reduce carbon emissions?" *Environmental and Resource Economics*, 29(4), 379–400.

BP Statistical Review of World Energy 2018, https://www.bp.com/content/dam/bp/en/corporate/pdf/energy-economics/statistical-review/bp-stats-review-2018-full-report.pdf.

Copeland, A. (2009). "Commodity outlook: Natural gas," *Australian Commodities*, 16(4), 658–662.

https://en.wikipedia.org/wiki/World_energy_consumption

https://www.hydrocarbons-technology.com/wp-content/uploads/static-progressive/nri/hydrocarbons/Projects/Roadrunner/roadrunner-l.jpg.

Paliwal, P. (2010). "City gas India roundtable 2010: Initiatives and challenges," *Vikalpa-The Journal for Decision Makers*, 35(4), 61–91.

Paliwal, P. (2017). "Natural gas pricing," in *Natural Gas Markets in India-Opportunities and Challenges*, Kar, S. K., and Gupta, A. (Eds.), Springer, Singapore.

www.naturalgas.org

# 2

# *Regulatory Aspects of Natural Gas Transmission and Distribution Business*

## 2.1 Introduction

Natural gas transmission and distribution is an important activity in the natural gas value chain that concerns a variety of stakeholders. Although producers/marketers, and customers are indeed major stakeholders, the fact that pipeline networks (both transmission and distribution) affects not only these two stakeholders but also the public, environment, and government—entities that are also justifiable stakeholders. Natural gas transmission and distribution must be viewed in its entirety and needs an adequate governance framework that caters to the interests of all these stakeholders—a need that has been evident worldwide for quite some time. Various natural gas markets worldwide, depending upon their stage of maturity, have from time to time created such governance frameworks. Although these governance frameworks—expressed largely by a set of regulations—may be different in content, they are broadly same in intent. The intent being balancing the interests of all stakeholders, ensuring transparency, and facilitating the growth and development of natural gas markets. This chapter discusses all concerned regulations dealing with different aspects of natural gas transmission and distribution. It would be contextual to mention that natural gas exploration and production (E&P) is also guided by a distinct set of regulations in all markets worldwide. However, given the scope of our discussion, we will adhere to the regulations related to the transmission and distribution part of the natural gas value chain.

## 2.2 Role of Regulator-Balancing Act

Any regulator must balance the interest and expectations of all stakeholders—largely consumers and operators/marketers. And maintaining this balance is the essence of natural gas regulations.

### 2.2.1 Consumer Interests and Expectations

**Low tariff:** Consumers in general will look for tariffs which are not exorbitant. That means they would aim at keeping their energy (in this case natural gas) bills as low as possible. Any rational consumer can be expected to express this expectation and justifiably so.

**Choice of supplier:** It is natural for all consumers to have their own say in choosing their natural suppliers/marketers. Although depending upon the stage of natural gas market development, this choice may or may not be always possible (or may be possible at later stages). However, the expectations are not unusual. That is because consumer interests are best addressed when they have freedom to choose.

**Quick access:** Consumers always prefer avoiding waiting times for access to any form of any fuel supplied by energy utilities and the natural gas business is not an exception.

**No upfront deposits:** To ensure that their cash outflow in accessing the services of an energy utility are as low as possible, consumers would prefer not parting with any deposits. Although this preference may be contrary to the expectations of the utility companies, consumers indeed expect this arrangement as they feel it also reduces their (future) switching costs, if any.

### 2.2.2 Interest and Expectations of Natural Gas Marketers

**Tariff providing adequate returns:** Natural gas marketers have a rational expectation of getting a fair return on their investments that they have made in creating time-consuming and capital-intensive transmission and distribution networks.

**Exclusivity:** Having invested in either of the businesses (transmission or distribution), the operators here would have an expectation about protecting their interests by allowing them to enjoy exclusivity for operating in their networks and markets. However, this expectation of exclusivity also has been one of the greatly debated issues among regulators and other stakeholders (marketers and customers included) and different regulatory bodies have dealt with this expectation in different ways.

**Phased implementation of network:** Natural gas transmission and distribution businesses are evolving worldwide. Certain pioneer markets, such as in the U.K., have reached a stage where the markets have become mature and predictable to a large extent. However, many markets are in various stages of development. As mentioned previously, getting into either one of the businesses is highly capital intensive. If capacities once created remain unutilized (for whatsoever reasons), then it is a drain on the operators' time and capital.

Thus, the operators have a genuine expectation of being allowed to roll out their networks gradually in synchronization with the market development.

**Access based on economic viability:** Governments, in general, would like to ensure that all segments of the markets, irrespective of their economic status, have an equal opportunity to access the services of natural gas utilities. Thus, they expect the roll out of networks to happen in all parts of concerned geographical areas. However, utilities at times would expect the freedom to lay networks (particularly distribution networks) according to their estimation of economic viability of sub-parts of the geographical areas and markets. This estimation means they would like to serve those sub-markets first that have a demographics which could not only afford natural gas as a fuel but also consume relatively large volumes. Other sub-markets would be a secondary or tertiary priority of the operators. However, many governments consider this approach as "cherry picking" which contradicts the concept of Universal Service Obligation (USO) also known as Market Service Obligation (MSO).

**Customer accountability and creditworthiness:** Continuing with the expectation of access based on economic viability, the operators and marketers would like a regulatory regime that makes customers equally responsible in terms of drawing of committed natural gas, ensuring non-defaulter of bill payments, and so on.

## 2.3 Objectives of Natural Gas Regulators

As mentioned previously, the role of a natural gas regulator is to ensure an adequate balance between varying (and at times conflicting) interests of different stakeholders. Major objectives of any natural gas regulator are as follows:

**Promote competition:** Competition ensures efficiency and best deals for consumers. It also ensures that no player is ever able to use market dominance based on exclusivity to further its own interests. Thus, even if exclusivity must be provided to any player (depending upon market peculiarities), it must be in a transparent framework designed by the regulator.

**Maintain or increase supplies:** Regulations are framed with an objective to see that adequate quantities of natural gas according to demand are made available on a consistent basis. Thus, many regulations have both incentive and penal provisions for operators and marketers.

**Protect end-consumers' and investors' interests:** Regulator works as not only as a framer of the rules, but also as an arbitrator in case of intervention to protect the interest of consumers. This process is according to the law of natural justice, where unorganized consumers (who have the least access to any judicial framework) do not get a raw deal at the hands of operators and players.

**Avoid duplicate investments:** As mentioned previously, it is in the best interest economically that capital-intensive natural gas transmission and distribution networks work on an adequate capacity utilization basis. If their installed capacity remains unutilized, then it is an economic waste for all concerned. Thus, the regulators must ensure that unnecessary and redundant capacities duplicating the existing capacities are not created in the market. Any addition in capacity should be allowed only when existing capacities are unable to meet the demand and supply needs.

**Define and enforce quality parameters:** Natural gas transmission and distribution systems are complex technical activities. Moreover, they also have health, safety, hazard, and environmental implications. Thus, a structured and standardized approach is necessary to ensure all aspects—right from laying down pipelines, pipeline material, gas flow, and operation and maintenance (O&M)—are carried out with utmost adherence to quality and reliability. It is an important objective to define and ensure implementation of all these standards.

Apart from these issues, the regulators also play an important initial role in inviting all prospective players to build, own, and operate the natural gas transmission and/or distribution networks. That means no entity without being registered with and authorized by the regulator may operate these businesses. The natural gas regulator has a thorough process in place to determine the capability and commitment of these operators and marketers.

## 2.4 Drivers of Regulatory Decisions

Having understood the varying interests of stakeholders and subsequently the role and objective of natural gas regulators, it would be now contextual to know what drives the decisions taken by natural gas regulators. That is because irrespective of the varying interests of stakeholders and regulator objectives, decisions also need to come from a broader perspective that helps the overall long-term natural gas market development. These drivers can be broadly categorized as follows:

**Stage of market development:** Regulators' decisions are normally based on the different stages of natural gas markets (i.e., either mature, under transition, or nascent). That is because the needs, expectations, and interests of various stakeholders are different in different stages of market development. A rigid, unified approach may not be suitable for different market stages.

**Transparent principles (win-win for entity and consumers):** All decisions must be driven by the principles of transparency and that it's not a zero-sum game. This principle drives decisions by natural gas regulators that are optimal and have economic and market development rationale.

**Learning from other sectors and other countries:** Regulators' decisions are also largely driven by experiences from other sectors (like telecommunications, solar energy, insurance, and so on) that also operate under a well-defined regulatory framework. Also, they learn from similar experiences of natural gas regulators from other countries. For example, natural gas regulators in an evolving market like India (which has a single digit share of natural gas in its energy basket) must learn more from similar market abroad.

## 2.5 Regulations: An Overview

Major regulations for cross-country transmission pipelines and city gas distribution (CGD) networks can be broadly categorized as follows:

- Authorization by choosing entities through a transparent bidding process
- Marketing and infrastructure exclusivity for CGD
- CGD network/compressed natural gas (CNG) compression/pipeline tariff
- Quality of service
- Technical and safety standards
- Integrity management for CGD networks and cross-country transmission pipelines
- Operating code

### 2.5.1 Authorization by Choosing Entities Through a Transparent Bidding Process

Keeping in view long-term development of natural gas markets, it is important to have serious players in the distribution business. These players need to be financially sound and align with the regulators' objectives of a balanced market development with offering competitive tariffs (and natural gas prices wherever applicable). These prospective players also must be prepared to invest in building ample infrastructure.

Generally, this process is ensured by judging the bidders on the following parameters:

1. Offering competitive gas transmission and distribution network tariffs and CNG compression charges and pipeline transmission tariffs over the economic life of the project.
2. Commitment towards building pipeline infrastructure.
3. In case of CGD networks, *ceteris paribus*,[1] commitment towards extending the maximum number of domestic connections gets a considerable weight over other bidders extending less. This criterion is to exhibit the seriousness on the part of bidder towards the balanced development of CGD market, because, generally, it is cumbersome to serve domestic connections due to their large numbers, geographical spread, and very low intake of piped natural gas.
4. Laying out of network in phased manner as stipulated by the regulator.

### 2.5.2 City Gas Distribution Network, Compressed Natural Gas Compression, and Pipeline Tariff

To ensure a fair deal to the CGD network, the regulator aims at keeping the network tariff (the tariff charged by CGD company to pass natural gas through the CGD network) competitive. Although the regulator should not aim at fixing the tariff, the tariff should, through adequate regulations, always facilitate inviting bids in such a way that, *ceteris paribus*, the most competitive tariff bidder wins the award. The same principle is also applicable to operators bidding for laying cross-country pipeline networks.

However, the process certainly does not mean that the bidder must compromise financial feasibility for business in ensuring the competitive tariff. Towards this end, the regulator develops a mechanism wherein both "a reasonable rate of return" and "a competitive network tariff" are balanced. The bidder is asked to share the data regarding the methodology

---

[1] *Ceteris paribus* is a Latin phrase used in economic studies which means "all other things being equal." This phrase has significance throughout the discussions on regulations.

used for calculating network tariff and compression charges for the CNG/transmission pipeline tariff, capital employed, cost of capital, capital expenditure details, operation and maintenance (O&M) costs, manpower costs, depreciation formulas, and other financial costs such as insurance *premia* on fixed assets and so on.

The regulator has exclusive rights to reject the bid if, despite feasibility problems, an entity is providing low network tariffs, compression charges, and transmission tariffs because that may make the project vulnerable at a later stage. At the same time, the regulator also may ask the bidder to rework the costs as well as network tariff, compression charges and transmission tariff, if it is found that the bidder has inflated certain costs.

The regulator in a transparent manner defines in advance what (assets and equipment) may be considered under capital expenditure so that there is a common rule to calculate return on capital employed. Similarly, depreciation methods also are well defined and given in advance.

### 2.5.3 Marketing and Infrastructure Exclusivity for City Gas Distribution

As discussed previously, CGD entities in developing or nascent markets have an expectation to operate with exclusivity—at least during the initial few years of their business—so that they can get a fair return on their investment. They consider this privilege an inventive for the business risk they have taken in a relatively uncertain natural gas distribution business. Such expectations deal with concerned stages of market development in respective markets. Many of the matured markets today were evolving a few years back and, hence, the regulators there have dealt with such expectations of exclusivity based on the market conditions prevailing at the time.

Even now, regulators in some markets have deliberated on this exclusivity expectation and have offered a rationale wherever granting such exclusivity. Following is the case study of the Indian CGD market where the natural gas regulator called Petroleum and Natural Gas Regulatory Board (PNGRB) has carefully considered the aspect of exclusivity and has offered the rationale for allowing exclusivity to entities that lay, build, operate, or expand a CGD network as follows:

> *Exclusivity for laying, building, or expansion of the CGD network during the economic life of the project; and in terms of an exemption from the purview of the contract carriage or common carrier for a limited period of time- is envisaged with a view to facilitate the development of a planned and integrated CGD network with appropriate priorities for end-use of natural gas as also the network spread besides providing incentive to the entity for investing in such project.*
>
> *Exclusivity is deemed necessary to facilitate the development and operation of an integrated network by a single entity as per the prescribed technical standards, specifications including safety standards. This shall also preclude multiple digging-up of lanes, roads, etc. in the authorized area.*

*Exclusivity as per PNGRB is also deemed to be necessary due to the following reasons, namely:*

*a) During the initial phase of the development of city or local natural gas distribution network, there is a need to have a close synchronization between the development of requisite infrastructure and the ramp-up in the natural gas volumes for different end-consumers in different areas. It is expected that the development of the city or local natural gas distribution network would be quicker if the same is guided by entity's own plan (which is responsible for meeting various service obligations) rather than the expectation of other potential marketers of natural gas in the network. Also, it is more practical for the regulator to deal with one entity rather than multiple entities to ensure a strict compliance with the service obligations by the entity in the initial period;*

*b) During such limited period of exclusivity, the authorized entity could be made directly responsible for meeting the desired service obligations, namely, achieving maximum PNG (Piped Natural Gas) domestic connections and other related aspects;*

*c) Besides such an approach is also likely to incentivize investments in this capital-intensive business.*

*Ideally, while the exclusivity is to be for the economic life of the project (normally considered to be 25 years), the exclusivity depend upon various factors, namely, the projected natural gas demand build-up in the city or local area (which in turn would depend upon the key drivers for demand in that city or local area, such as, level of industrial or commercial activity, vehicular population and conversion of vehicles in to CNG, potential domestic PNG customers, consumer preferences, price of alternative fuels, etc.), geographical spread and population, projected capital cost of the project, investment climate, etc. However, considering that these factors vary from city to city, a credible assessment of exclusivity period based on these factors may not be always practical. Thus, normally the period of exclusivity may be limited to initial few years for cases where an entity proposes to lay, build, operate or expand a CGD network.*

*(Source: Petroleum & Natural Gas Regulatory Board (PNGRB), Government of India; http://www.pngrb.gov.in/OurRegulation/CGD-Network-GSR198.html.)*

The regulator, looking into other factors at the expiration of this exclusivity period, may make an appropriate decision about extending the exclusivity period or otherwise.

In a similar spirit, apart from marketing exclusivity, the regulators also decide infrastructure exclusivity for certain periods. This process considers the fact that while an entity builds a CGD network and is a marketer, it should be allowed to own and operate infrastructure (CGD network) even after the marketing exclusivity period is over. In that case, another marketer is liable to pay requisite tariffs to the owner of the network.

### 2.5.4 Quality of Service

It is important that CGD network customers receive an uninterrupted supply of natural gas from the distribution companies. In absence of this supply, they may be put into unnecessary hardships. If left up to distribution companies, it is possible that they may attempt to avoid their responsibility. Thus, it becomes incumbent upon the regulator to protect the interests of consumers and formulate relevant regulations. These regulations deal with norms regarding providing consistent supply of natural gas and requisite services to consumers. Apart from these norms, the regulations also deal with reciprocating obligations of consumers—thus also taking care of the distribution companies' interests.

### 2.5.5 Technical and Safety Standards

Technical and safety-related regulations are applicable to ensure that the concerned definitions, design, materials and equipment, welding, fabrication, installation, testing, O&M corrosion control, modifications, and abandonment of transmission network/CGD network/CNG stations are in accordance with requirements of international standards (like ASME, ISO 9000). Technical and safety regulations also are intended for developing standards to ensure a standardized application of design principles and to provide guidelines for selection and application of materials, equipment, and components.

Safety regulations also ensure that the CGD system has a leak detection system in place that is operational. Further, safety regulations ascertain that the transmission pipelines, CGD system, and CNG stations have in place the necessary pressure relief valves and that there is protection of the system against third party damages (to common public) both in respect to carbon steel pipe, steel pipe, polyethylene pipe, and copper tubing. Other areas tackled by the safety regulations also include corrosion control in pipelines by using sophisticated techniques such as cathodic protection.

These regulations also address aspects such as design, fabrication, installation, testing at the time of construction, and commissioning. Sometimes, a third-party audit (TPA) is carried out to ascertain the compliance of these regulations. Regulations clearly outline how to conduct business with TPAs. Relevant non-destructive testing (NDT) methods are required to be used to ascertain quality and workability of pipeline infrastructure in a transmission network as well as in a CGD network. And the regulations specify all these requirements.

In the distribution network, the technical and safety regulations further extend to the CNG station (CNG mother station[2]/CNG on-line station)* and CNG daughter station* which are required to be designed, operated, and maintained in line with the requirements of the prevailing safety standards related to handling explosives. This process includes compression, handling, and transportation activities of CNG.

### 2.5.6 Integrity Management for City Gas Distribution Networks and Cross-Country Transmission Pipelines

Transmission networks as well as CGD networks have important assets for transporting combustible gas under pressure at long distances as well as within the densely inhabited areas. As such, they put people, communities, and the environment at risks in cases of malfunction. On the other hand, cross-country transmission pipelines as well as CGD networks are also vulnerable to external damages caused by third parties and in several cases such external damages may result in network failures. In case of failure, many parties may be affected. Therefore, introduction of a robust system which ensures maximum availability of both networks with minimum interruption and damages is vital. It is, thus, the role of the regulator to codify, communicate, and execute a suitable integrity management system (IMS) for transmission as well as for CGD networks.

An IMS for cross-country pipelines and CGD networks provides a comprehensive and structured framework for assessment of transmission pipelines and CGD network conditions, likely threats, risks assessment, and mitigation actions to ensure safe and incident-free operation of transmission and CGD networks.

Regulations ensure that an effective IMS contains the following features:

1. Ensuring the quality of transmission pipeline and CGD network integrity in all areas which have potential for adverse consequences.
2. Promoting a meticulous and methodical management of transmission pipelines and CGD network integrity and mitigate the risk.
3. Increasing the general confidence and faith of the public in operation of transmission pipelines and the CGD network.

The intent of regulations is to optimize the life of the transmission system and CGD network with the inbuilt incident implementation of Integrity Management Plan (IMP) investigation and data collection including review by the entity.

---

[2] Different types of CNG stations. These shall be discussed in detail further.

Relevant regulations ensure that such a comprehensive IMS has the following features:

1. **Planning for integrity management:** This includes gathering and substantiation of data, evaluation of gamut of risks, classification of risk, appraisal of integrity with reference to risks, risks mitigation, updating of relevant data, and re-evaluation of risk.
2. **Performance appraisal of the integrity management plan:** This process scrutinizes the efficacy of the IMP and provides inputs for improving it further for best results.
3. **Communication plan:** This plan pertains to dissemination of relevant information with a smooth, timely, and transparent flow to all the stakeholders who may be affected with the compromise in integrity of pipeline systems—transmission as well as distribution networks.
4. **Managing change:** This process deals with bringing about a significant cultural change in respect to the attitude towards risks and dealing with them.
5. **Quality assurance:** This process deals with defining quality, planning, and executing processes that ensure quality at all stages in implementation of IMS.

Regulators should ascertain that an IMS is designed with the objective of ensuring the integrity of transmission networks and CGD networks with almost zero downtime to ensure no damage to environment, and minimizing business risks related to the operation of gas networks. The availability of a well-designed IMS facilitates transmission and CGD personnel in integrity tasks to ascertain work plans and goals in different time horizons which in turn improve their efficiency and professional achievement.

Although the IMS facilitates the transmission network and CGD operators in selecting an identified system for implementation such that the IMS is standardized for all transmission networks and CGD entities within the specified market, it is the job of the regulator to codify it and ensure its implementation.

### 2.5.7 Operating Code

The cross-country transmission network is meant to supply gas to multiple destinations according to requirements. Sometimes the transmission network operators may carry their own natural gas for delivery to their own customers. Other times they may act merely as transporters of the natural gas owned by other entities (known as shippers) and deliver gas according to the instructions of the owners of the commodity. In such cases, the transmission network owner receives the transportation tariff.

Looking into the dynamic relationship between various entities, regulators in many markets have developed a set of regulations that guide the commercial relationships between these entities under various operating conditions. Such regulations are known as operating code regulations.

Regulations regarding operating codes are intended to document certain rights and responsibilities of each of the parties involved in the transportation of gas through the transmission pipelines. As mentioned, the transmission of gas may be from multiple sources to multiple consumption points. This operating code of conduct addresses technical and operational issues relating to transfer of gas through the transporter facilities.

The operating code sets out the generic and specific guiding principles that govern the transporters' and shippers' respective rights and obligations in all aspects related to usage of transmission pipelines for the transportation and flow of gas in requisite quantities and specifications. The operating code is intended to provide a clear, fair, transparent, and a non-discriminatory framework for shippers wishing to use the transporter facilities and for transportation of the transporter's own gas through the transmission network.

## Bibliography

Paliwal, P. (2011). "Legal tangle in institutional selling: Case of gas distribution company," *Oil, Gas and Energy Law Intelligence*, 9(5).

https://www.econlib.org/library/Enc/NaturalGasMarketsandRegulation.html

www.energy.gov

www.iea.org

www.pngrb.gov.in

Zwart, G. (2009). "European natural gas markets: Resource constraints and market power," *Energy Journal*, Special Issue, 151–165.

# 3

# *Natural Gas Transmission Business: Project Management Aspects*

The efficient and effective movement of natural gas from producing regions to consumption regions requires an extensive and elaborate transportation system because producing regions may be far away in the remote locations. Therefore, sometimes natural gas produced from a specific well may need to travel a great distance to reach the point of use. The transportation system for natural gas consists of complex networks of pipelines designed to quickly and efficiently transport natural gas from its origin (upstream) to areas of high demand for natural gas (downstream).

Natural gas transmission business refers to a business wherein the firm transports natural gas from source to market. The source can be an oil and gas field from which the natural gas is extracted through exploration and production (E&P) activity or re-gasified liquid natural gas (RLNG) from liquefied natural gas (LNG) terminals. The market is the consuming center which consists of compressed natural gas (CNG) for vehicle users, piped natural gas (PNG) for domestic users, commercial users, and industrial users. The commercial users and industrial users comprise a major chunk of the business. The transmission business requires firms to lay and operate the pipeline which is offered to users (natural gas distribution companies) on a common carrier principle regulated by the government or a regulatory authority.

Pipelines can be characterized as cross-country pipelines. Interstate pipelines are like an interstate highway system. They carry natural gas across state boundaries and, in some cases, across the country. Intrastate pipelines, on the other hand, transport natural gas within a specific state.

This chapter discusses the project management aspects of natural gas transmission pipeline business.

## 3.1 Natural Gas Transmission Pipeline Projects

The federal government usually plans natural gas transmission pipeline projects to connect the source to the market to ensure an energy supply to users. Based on studies related to the availability of natural gas at the source and demand for natural gas in the market, the government invites bids from the market players to implement and operate the transmission line project. The interested firms based on their strategic analysis of the government policy and regulations plan to participate in the bidding process. The participating firm undertakes a techno-economic feasibility study to assess the profitability potential of the project.

While participating in the bidding process, the firm evaluates the bidding criteria for competitive bidding. The criteria, in general, used by the various governments and regulatory authorities to evaluate the bids follow:

1. Technical Bid Evaluation Criteria: Technical experience in operating and managing the natural gas transmission lines, adequate technical competency of personnel, net worth of the bidder, and so on.
2. Financial and Commercial Bid Evaluation Criteria: Evaluation of financial bids is based on the present value of yearly tariffs quoted for different tariff zones (zones are based on line kilometers from source to users) and volume to be transported (zone-wide) for the economic life of the project (i.e., 25 or 40 years).

Sometimes the technical bid may be evaluated only to check the eligibility criteria. To award the project usually the financial or economic bid is evaluated. To evaluate the bids, weights are assigned to various parameters and the bidder with the highest composite score is declared as successful for the award of the project. The qualified bidder is then permitted to implement the project.

## 3.2 Natural Gas Transmission Pipeline System Description

The natural gas transmission pipeline system delivers natural gas from the source through regulating stations to dispatch stations. The send-out gas at dispatch stations will be at a controlled temperature and the pressure of send-out gas needs to be less than or equal to the maximum operating pressure (MOP). Furthermore, the send-out gas pressure depends upon downstream consumption and pipeline conditions. The design pressure needs to be greater than or equal to MOP. The system has dispatch and receiving terminals. The facilities include an emergency shut down (ESD) valve system to isolate a pipeline section when necessary. It also includes filtration, metering, pig launcher, and pig receiver facilities along with related valves. It also has a facility to depressurize the pipeline if required. Multiple intermediate pigging (IP) stations at about 100 kilometers apart are installed with pig launcher-pig receiver and related valves to manage pigging operations. The IP station also can be used, if required, to depressurize the pipeline. In between IP stations, multiple sectionalizing valve (SV) stations are installed at intervals of 20–25 kilometers. The SV stations include both SV remote (SVR) and SV manual (SVM). Compressors with valves for by-pass arrangements at remote-operated SV stations are installed. Compressors are used to ensure required flow rate and pressure at the user's location (downstream) in case the upstream pressure is less than required.

The bi-directional pipeline systems are planned for future requirement with the characteristic for flow in both directions. In this case, the pig receiver will become a pig launcher and vice-versa.

Figure 3.1 shows the natural gas transmission pipeline schematic diagram. The various elements of the gas transmission system are briefly discussed next.

### 3.2.1 Receiving Stations and Dispatch Stations

A receiving station has filtration, metering, pig receiving, and pig launching facilities with related valves and a facility to depressurize the pipeline when required.

A dispatch station has a flange, a main inlet valve (ball valve) to isolate the facility if required, a filtration skid, a metering skid, a pig trap, a by-pass valve to temporarily switch off the inlet when pig is to be launched, an electric power supply with an uninterruptible power source (UPS), and instrumentation and supervisory control and data acquisition (SCADA) panels for connection of the pipeline system to the SCADA system, lighting, fencing, and security cabin (Figures 3.2 through 3.3).

### 3.2.2 Intermediate Pigging Stations

Pigging operations refer to the activity of cleaning the pipeline periodically or inspecting changes in the pipeline features due to corrosion and other conditions using the pigs. Pig refers to the pipeline inspection gauge usually used in piggable lines. The frequency of pigging is determined based on the quantity of debris found during pigging. Pigging cleans the pipeline so that the capacity and life of pipeline is maximized.

IP stations contain the pig launching (PL) and pig receiving (PR) facilities. They are installed at an average of about every 100 kilometers. As mentioned previously, a pig launcher can also function as a pig receiver in bi-directional pipeline. PL and PR facilities also are installed at dispatch and receiving terminals and they are manned.

All IP stations have electric power supply with UPS, instrumentation, and SCADA panels connected to the SCADA system, lighting, fencing/boundary walls, and security cabin.

### 3.2.3 Sectionalizing Valve Stations

The SV stations consist of valves of the same diameter (also referred as mainline valves) as those of the pipeline along with some associated station piping and fittings such as flanges, elbows, other small diameter valves.

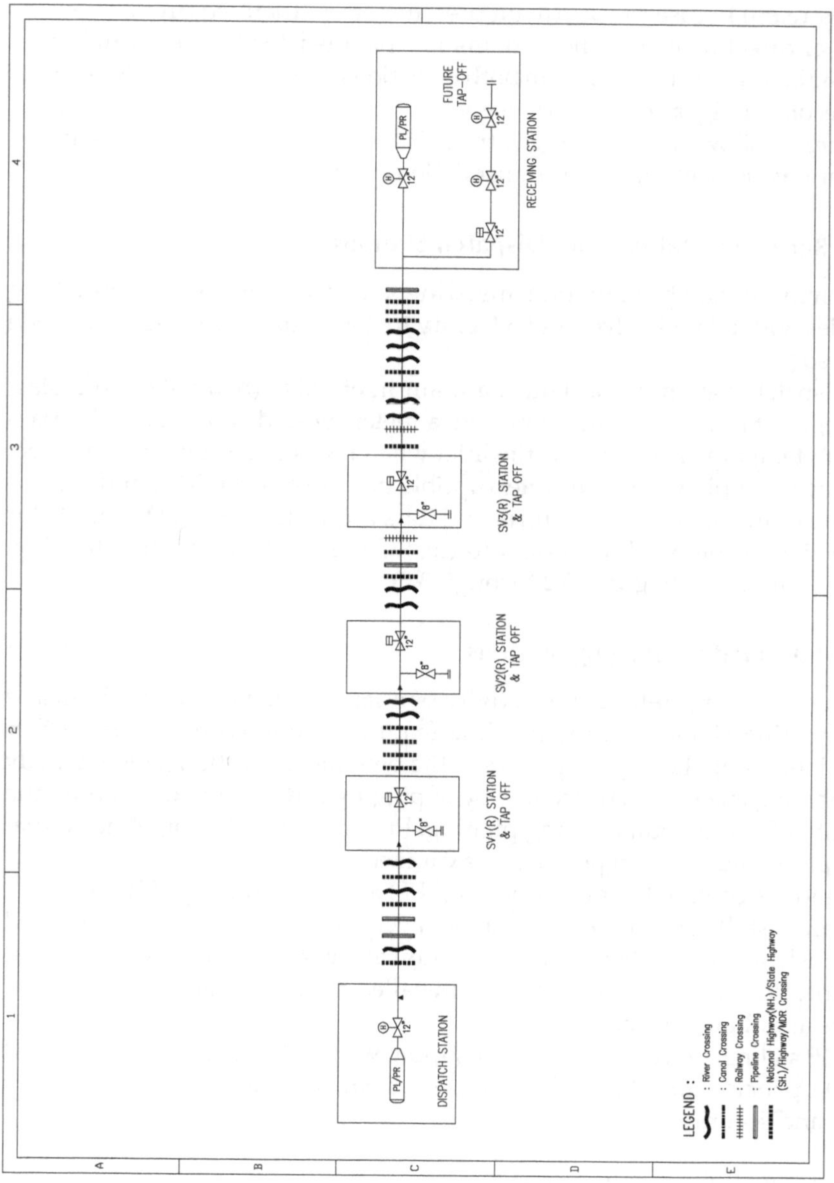


**FIGURE 3.1**
Schematic diagram of a natural gas transmission pipeline system.

**FIGURE 3.2**
Station piping and crossovers inside the process area. (From www.gspcgroup.com.)

**FIGURE 3.3**
Station piping with crossover and instrumentation (i.e., pressure gauge). (From www.gspcgroup.com.)

**FIGURE 3.4**
Mainline actuated valve at SV stations. (From www.gspcgroup.com.)

These mainline valves are meant for isolation of pipeline sections, if required, during leakages or maintenance work (Figure 3.4). The SV stations are generally unmanned. Usually, these SV stations are built 20–25 kilometers apart on average. Also, SV stations may contain tap-off in case if any spur line is to be connected in the future to serve prospective customers. The SV valves can be marked to identify them because they are installed underground.

The SV station consists of a flow tee for by-pass, the main pipeline with valves (which can be operated from central location through SCADA), instrumentation, SCADA for SVR, and electric power supply for local instrumentation, lighting, fencing, and security.

The by-pass consists of a ball valve and a throttling valve, to pressurize at the time of commissioning and re-pressurize when needed, for controlled blow down and throttle adequately for the flow during pigging.

### 3.2.4 Compressor Stations and Sectionalizing Valves

Compressors are generally required when the line pressure is much less than operational requirement. This situation may arise in the case of very long pipelines passing through many states. Compressors are also required when the pressure at the source gets depleted and demand at the location of the user increases. The compressor station acts as a SV station and a compressor station. It has a compressor to add required pressure in the pipeline, a facility for pigging operations, and valves with local control and remotely control by a central control room equipped with SCADA. It also has by-pass with valves in between (Figure 3.5).

**FIGURE 3.5**
Compressor station. (Courtesy of BOTAS, Ankara, Turkey.)

### 3.2.5 Tap-Off Stations

The purpose of a tap-off station is to act as a SV station and a tap-off location (i.e., connection to the natural gas distribution company, the gas utility) for future pipelines based on the demand in adjacent areas of the pipeline route. It must have a flow tee for a future pipeline, the main pipeline with valves to control the flow, and a by-pass line.

### 3.2.6 Blow Down

As such the facilities for operational and emergency pipeline depressurization exist at all pipeline ends and IP stations. At pipeline end and IP stations the upstream or downstream part can only be vented by a single valve at a suitable location. However, at SV stations, venting can be done for both upstream and downstream by providing two ball valves.

In the transmission pipeline system, all the operating valves more than 12 inches in diameter are equipped with an actuator to assist the manual operations from a local panel or allow remote control operations. It is tough to operate valves more than 12 inches in diameter by hand and it also takes long time. All equipment facilities at SV stations and tap-off stations are buried except for instrumentation, actuators, and the vent stack. All the facilities at dispatch and receiving main stations are above ground.

## 3.3 Transmission Line Project Planning

The natural gas transmission pipeline projects are planned based on the demand of natural gas on the route of the pipeline. Since the design life of the pipelines is about 40 years, the demand it will serve for next 40 years on the route to nearby geographical areas is considered for the project analysis. The demand for all four segments (i.e., industrial, commercial, CNG, and PNG) that will be served across the pipeline route is to be forecast to plan execution of the project. These consumers will be connected through the spur line from the trunk lines or transmission lines.

When planning the transmission pipeline projects, the following decisions related to pipeline optimization are studied and analyzed:

1. Decisions related to optimal pipeline layout and route optimization (network topology)
2. Decisions related to optimal diameter of various pipes
3. Decisions related to the locations and number of valves, regulators, and compressors (collectively referred as machines)

### 3.3.1 Route Optimization (Pipeline Route Planning)

Steps involved in pipeline route selection are (1) desktop study (with secondary data), (2) reconnaissance survey (ground inspection), (3) finalization of pipeline route, (4) detailed engineering survey, and (5) generation of alignment sheets and cross-section drawing.

The desktop study involves identifying the nature of the route in terms of major crossings, terrain, and type of land along the route (i.e., agriculture, rocky, rugged, and so on). The route is split into chainages of about 50 kilometers with location identity. While planning the design for maximum ground temperature for hydraulic calculations, the maximum ambient air condition for compressor turbine sizing is considered in line with the climate of the country where the project is to be implemented.

The reconnaissance survey for the proposed route helps in preparing the detailed feasibility report (DFR). The proposed route and demand across the route are used to prepare the DFR. The tap-off locations along the route are planned based on analysis of the market survey report.

The reconnaissance survey helps in identifying multiple layouts to connect the supply and demand points. The objective is to select the shortest feasible path from the supply source to the demand source. As such, the shortest route may not be optimal because of the varying demand of customers. Therefore, the solution to this problem may be connecting the main trunk line with the spur lines to connect to the demand centers.

Approaches used so far for layout and route optimizations are (1) metaheuristics (genetic algorithm, simulated annealing, and dynamic programming), (2) non-linear optimization, and (3) integer programming

#### *3.3.1.1 Criteria and Guidelines for Route Selection*

Consider all the following criteria and guidelines simultaneously for route selection:

1. Select shortest feasible route
2. Avoid mining and defense establishments
3. Avoid any protected areas as far as possible
4. Avoid wild life sanctuaries, inhibited area, and so on
5. Avoid unstable ground features
6. Select the route with easy and favorable terrain conditions free of large water bodies
7. Avoid rocky, marshy, and low-lying areas
8. Avoid densely populated area
9. Select the route with easy approachability and accessibility for construction and maintenance
10. Avoid rail, roads (i.e., national and state highways), rivers, and canal crossing
11. Utilize existing right of way (RoW) and right of use (RoU)
12. Select favorable ground profile
13. Consider operability, maintainability, and augmentation capability

A few other important factors are:

1. Safety of public life and property
2. Environmental impact and avoid environmentally sensitive areas
3. Existing and future developments in the region such as roads, rail lines, canal network, reservoirs, townships, and industrial units which may be avoided
4. Avoid rugged and intricate grounds with hard strata, exposed rocks, boulders, and quarries

All the available topographic sheets of the area where the pipeline will pass through are made available for reconnaissance survey. The ground profile chart is made for the route.

The constraints in routing are studied. As such the layout problem is mostly seen in isolation; therefore, it is difficult to find the optimal solution. The better way to address this problem is to use analytical hierarchy process (AHP) approach through which we can assign some weight to all factors and thus have an overall insight to get a better solution.

#### 3.3.1.2 Characteristics of the Analytical Hierarchy Process Approach

The AHP approach (1) is a subjective approach with necessary objectivity, (2) uses a multiple criteria decision-making methodology, and (3) requires active participation of decision makers in reaching agreements and a rational basis for decision making.

#### 3.3.1.3 Route Selection Process Using the Analytical Hierarchy Process Methodology

The steps for the route selection process using AHP methodology follow:

1. Identify all alternative routes
2. Prepare a strong database for each route
3. Identify factors and sub-factors leading to route selection
4. Formulate risk structure in line with AHP requirements
5. Compare factors and sub-factors (pair wise—route comparison) to determine importance of these factors for selecting a route
6. Select alternatives with respect to each of the factors and sub-factors to determine the preference for alternative route over others
7. Synthesize the results across a hierarchy to determine the optimal route

#### 3.3.1.4 Factors and Sub-factors Used for Route Selection

The factors and sub-factors used for route selection are:

1. Pipeline length
2. Operability, which is analyzed keeping in view the ground profile, expansion possibility, and route diversion
3. Maintainability in the context of corrosion and third-party activities
4. Approachability in terms of proximity to transport facility and terrain characteristics
5. Constructability with respect to statutory clearances, mobilization of resources, availability of water and power, ease of construction, and environment friendliness

#### *3.3.1.5 Decision Support System for Deciding Importance of Factors*

The database for deciding the importance of factors may contain the following details:

1. Length
2. Number of stations
3. Number of SV stations
4. Terrain details (normal, slushy, rocky, forest, river, or populated area)
5. Soil conditions
6. Statutory requirements

### 3.3.2 Optimization of Diameter of Pipes and Locations and Number of Valves, Regulators, and Compressors (Machines)

To determine the optimal diameter and locations of machines, network analysis for evaluation of flow rates and pressure at various nodes (demand centers) is carried out. Pipeline Studio software is used for network analysis to optimize the diameter of pipes and locations and number of valves, regulators, and compressors. It is the state-of-the-art hydraulic simulation tool that quickly and accurately performs steady state and transient analysis of single-phase fluid flow in the pipeline network. The software has the capability to analyze single-phase fluid flow and heat transfer in the pipeline network, and calculate and report response values for critical system variables (i.e., pressure, flow, density, quality, and temperature) throughout the pipeline network at specified time intervals. The software can model both simple and complex pipeline networks and may include pipeline equipment such as valves, compressors and pumps, regulators, and heaters. Pipeline element attributes such as pipe length, wall thickness, roughness, and elevations can be easily assigned. The software has a design library that consists of a comprehensive set of pipeline components (i.e., compressors, pumps, regulators, block valves, heaters, and coolers). The software also helps create emergency plans and fine tunes operating parameters. Supplies and deliveries provide pressure and flow boundary conditions for the model and provide model constraints.

The steady-state simulation provides hydraulic results under the assumption that mass and energy flows are in equilibrium at every point in the pipeline network. The simulation may be run independent of transient simulation. The simulation determines the value of thermal and hydraulic variables like pressure, flow, and temperature under steady-state conditions. The transient simulation determines the values of these variables under transient conditions.

The Pipeline Studio offers following solutions:

1. Design: Design of networks, line size, and capacities
2. Planning: Evaluate pipeline response to operational changes
3. Engineering: Determination of line, pump, and compressor size
4. Contingency Analysis: Simulations of upsets, unusual events, and the evaluation of recovery action (i.e., compressor station failure)
5. Operational Planning: Demand balancing and inventory analysis
6. Operational Analysis: Evaluation of alternate mode of operation
7. Strategic Planning and Investment Program Analysis: Determine network requirements for typically 5, 10, and 15 years
8. Staff Training: Pipeline hydraulics and operations training in safe environment

**ILLUSTRATION**

The efficient movement of natural gas from producing regions to consumption regions requires an extensive and complex network of pipelines. Due to frictional head losses during the transportation and overcoming occasional differences in height it continuously loses pressure and thus the speed. To keep the flow of the natural gas constant, generally compressor stations are employed along the pipelines at distances of 70–200 kilometers. Efficient designs and operation of these complex natural gas networks can decrease the huge initial investment and daily operating costs. The operations of natural-gas pipeline systems are characterized by inherent nonlinearities and numerous constraints. As a rule of thumb, 3–5 percent of the gas transported is burned to power the transportation of remaining amount.

The use of compressors in a pipeline is attractive only up to a certain demand beyond which the compression cost increases drastically due to the nonlinear relation between pressure and flow. Beyond that demand using loop lines or increasing the diameter of pipeline may be a better option.

As mentioned previously, the simulation software, Pipeline Studio, is used to analyze the effects of compressors in a pipeline by running steady-state simulations to show nonlinear relationship between pressure and flow and complexities involved in natural gas networks. Simulation for different supply demand scenario cases at different locations may be undertaken. It helps in decisions related to optimization of the diameter of pipeline and other machines (i.e., compressors, valves, regulators, and so on).

(*Continued*)

**ILLUSTRATION (Continued)**

The analysis using Pipeline Studio software was undertaken for one of the existing transmission lines of about 1,150 kilometers in length. The pipeline was designed considering the life of the project as 40 years. Almost 10 years since its operations no compressors have been installed in between this transmission line. As discussed previously, usually compressors are deployed at the distance between 70 and 200 kilometers. Analysis was undertaken to find size of the pipeline diameter to be used for this transmission line. It was undertaken using a certain level of demand and pressure requirements at various centers connected to this transmission line. Three scenarios were analyzed where demand with certain pressure at location C varied which required an increase in the pressure at two locations in between (i.e., locations A and B). The scenarios analyzed were to serve the demand of (1) 11 mmscmd, (2) 14 mmscmd, and (3) 15 mmscmd at location C while employing compressors at locations A and B to increase the pressure to meet the demand at location C and find the size of pipeline diameter.

**ANALYSIS FOR SCENARIO 1 (i.e., DEMAND TO BE SERVED AT LOCATION C IS 11 MMSCMD)**

To meet the 11 mmscmd demand at location C, if the firm would have deployed two compressors at locations A and B, it could have saved capital expenditure (CAPEX) in terms of savings in the installation of lower diameter pipelines. The simulation results showed that the two compressors (i.e., 2.657 megawatt (MW) at location A to increase pressure from 78.63 to 94.6 bar and 2.160 MW at location B to increase pressure from 64.95 to 69 bar) need to be installed to meet the 11-mmscmd demand.

The analysis assumed compressor efficiency at 70 percent. The required capacity of compressor which could have been used may be at 3 MW each (availability of compressors with required capacity in market). The cost of compressor would have been USD 700,000 (price at the time of analysis) per MW. The analysis considered the life of compressor as 15 years and capital cost per year of compressor would be USD 280000. To run the compressor natural gas may be used and it is assumed that 1 MW energy need 6,000 mmscmd of gas and assuming the price of gas at USD 0.20 per mmscmd, the annual cost of gas will be USD 1,850,000. Assuming the operating and maintenance cost at USD 0.005 per mmscmd leads to total annual maintenance cost of USD 30,000.

Thus, the total cost will be USD 280,000 + USD 1,850,000 + USD 30,000 which is about USD 2,160,000. Considering the cost of pipes in market

(*Continued*)

**ILLUSTRATION (Continued)**

at the time of study, it was found that the total savings in pipeline cost by using a smaller diameter would have been about USD 53,400,500.

### ANALYSIS FOR SCENARIO 2 (i.e., DEMAND TO BE SERVED AT LOCATION C IS 14 MMSCMD)

To serve 14-mmscmd demand at location C, instead of larger diameter pipes two compressors (i.e., 2.317 MW at location A to increase pressure from 78.63 to 92.42 bar and 18.471 MW at location B to increase pressure from 41.1 to 84.8 bar) are required to be installed. The analysis shown below discusses the savings in terms of CAPEX by installing smaller diameter pipes.

The analysis assumed compressor efficiency at 70 percent. The required capacity of compressors which could have been used may be 3 MW and 19 MW (based on availability in market). The cost (at the time of analysis) would have been USD 700,000 per MW for 3 MW, and USD 500,000 per MW for 19MW. Assuming the life of compressor as 15 years, capital cost per year of compressor comes to USD 775,000 and the operating cost will be USD 8,000,000 (i.e., energy cost and maintenance cost) as per the analysis shown in scenario 1. Thus, the total cost would have been USD 8,775,000. The resulted savings in pipe diameter would have been USD 53,000,000.

### ANALYSIS FOR SCENARIO 3 (i.e., DEMAND TO BE SERVED AT LOCATION C IS 15 MMSCMD)

In this scenario the analysis showed that the power requirement at location B increases from 18.47 MW to 35.27 MW. It was observed further that when the power requirement shoots up for increase in demand by 1 mmscmd, capital and operating cost will also shoot up significantly. Thus, the analysis suggested that it is better to have larger diameter pipeline or loop line. The analysis also showed that there is a nonlinear relationship between flow and fuel consumption. The cost increases rapidly beyond certain levels of flow.

### CONCLUSION AND SUGGESTIONS

Natural gas transmissions networks are very dynamic in nature and usually do not operate on a simple steady-state basis. The exact sources of future supply and delivery points of future demand cannot be ascertained at the time of design as the pipeline has a life of 40 years. Therefore, it must be designed considering future

(*Continued*)

**ILLUSTRATION (Continued)**

operations. The utilization of the existing gas grid, which was analyzed previously, was very low at the time of analysis; therefore, they did not use any compressors in line. It is designed for a higher future demand. It was built in phases; therefore, demand and supply was also considered in phases.

From a strategic decision point of view (i.e., pipeline designing), we need to have good forecasting of business in terms of demand. The optimal diameters can be decided only based on accurate demand forecast. Also, it seems a good ploy to have compressors in line as their cost is much less against the high cost of laying larger diameter pipelines. The compressors are a better option than laying larger diameter pipes or using loop lines for reducing pressure drop across pipeline because the cost of laying larger diameter pipes is very high. However, compressor use in pipelines seems to be attractive only up to a certain demand beyond which the compression cost increases drastically, which we have observed in the simulation cases. Beyond that demand, using loop lines or increasing the diameter seems better option.

Also, the decision for the compressor location is a strategic decision which is based on demand at various points downstream. It may be planned considering the future demand in downstream locations. It may be optimal to place it in between the lines and not at the nodes. Use of multistage compression may be better than single stage since it may reduce the power requirements.

The studied company mainly used the pipe diameter sizes in multiples of six (i.e., 6, 12, 18, 24, and 30 inches) in their gas grid. But the pipe diameter sizes are available in almost every 2-inch increment. Therefore, considering the forecasted demand, some other size in between may prove optimal, which can result in huge savings.

## 3.4 Designing and Scope of the Transmission Pipeline (Technical Aspects)

### 3.4.1 Pipeline and Related Machines/Equipments

#### *3.4.1.1 Broad Specifications*

The location of the source (dispatch station) and the destination (receiving station), gas properties given by hydraulic studies, and other installations in the pipeline network length from source to destination are the basis of design.

#### *3.4.1.1.1 Pipeline Design Parameters*

The parameters used for design are number of days and hours of operations, pipeline operating life (may be considered as 40 years), pipeline length, main pipeline diameter, internal coating (80 micron thickness epoxy coating), material of construction (carbon steel), sub-soil temperature below 1 meter of ground for the entire length of the pipeline, design temperature for buried section and above ground section, operating temperature, design pressure, MOP, corrosion allowance, gas supply condition at dispatch station (i.e., inlet temperature and inlet pressure), and pipeline laying depth. Based on the design parameters, pipeline wall thickness and grades are decided.

The hydraulic study is conducted based on pressures at dispatch station, tap-off points and receiving stations. The study also helps in identifying locations for compressor stations. The location of dispatch terminal, receiving station, and tap-off points are also considered for the design of natural gas transmission pipelines.

### 3.4.1.2 Ultimate Flowing Capacity

Maximum flowing capacity for design is estimated based on hydraulic study, compressor stations with capacity and locations, demand at tap-off points, and receiving stations.

### 3.4.1.3 Compressors and Intermediate Pigging Stations

The compressors details to be planned are related to type of compressor, compressor drive (gas turbine), fuel source, cooling method, and compressor duty (i.e., capacity, flow rate of fuel, suction/discharge pressure, and so on).

The other material required includes bulk material (i.e., for piping: assorted pipes, valves, flanges, fittings, long radius (LR) bends, flame accessories, and electrical and instrumentation facilities). The instrumentation material and equipment are personal computer (PC)-based systems and programmable logic controllers (PLCs), gas chromatograph, transmitters, pressure and temperature gauges, instrument cables and junction boxes, fire and gas detection system, and so on.

### 3.4.1.4 Pipeline Design

Based on an optimization study, the design parameters for consideration are (1) length in kilometers, (2) diameter of the pipe in inches, (3) design pressure in kilograms per square centimeters (kg/cm$^2$), (4) design temperature in Celsius, (5) corrosion allowance in millimetres (mm), and (6) material of construction and so on.

The pipeline used is carbon steel. Usually it is American Petroleum Institute (API) 5L longitudinally/helically submerged arc welded (LSAW/HSAW) steel pipes with Grades X60, X65, X70, and X80.

LSAW pipes are preferred for the uniform property throughout the pipe body and they achieve close dimensional tolerance for better and easier fit at the site to ensure faster construction. These pipes are easily cold expandable and have better weld orientation, which allows placement of the weld at the least stressed position. Cold and hot bends can be easily and informally achieved and the pipe also has good fatigue resistance. With LSAW pipes the reference points can be easily identified during intelligent pigging operations.

*3.4.1.4.1 Line Pipe Specifications*

The line pipe specifications are product specification level (PSL), API 5L suitable grade LSAW/HSAW pipes. Pipes of carbon steel are used for natural gas transmission projects.

The parameters to select the pipe specifications are maximum operating pressure, maximum operating temperature, suitable steel grade, length of pipeline (usually 12 meters), method of manufacture (LSAW/HSAW), wall thickness, corrosion allowance, yield strength, maximum yield strength and tensile strength as per PSL, weld efficiency, and so on.

The hydro-testing pressure test is conducted at a manufacturing facility for every pipe for a minimum of 15 seconds. The calibration of the gauge is undertaken at the start of each shift (Figure 3.6).

### *3.4.1.5 Corrosion Protection Coating*

**External coating:** It must be suitable to a varied temperature range according to the climate in the country of use. It must have other properties such as high integrity, resistance to ageing and degradation,

**FIGURE 3.6**
Transportation pipelines. (Courtesy of BOTAS, Ankara, Turkey.)

resistance to attack by micro organisms, resistance to soil stresses, ease of application, and resistance to impact and mechanical damage during transportation. Three layers of extruded polyethylene coating on the external surface is applied according to DIN-30670. The thickness of the coating must be a minimum of 2.5 mm for three-layer polyethylene(3LPE) or polypropylene (3LPP). The sequence of application is first a layer or coating of powder epoxy primer, second a coating of polymeric adhesive, and third a coating of extruded polyethylene (Figure 3.7).

**Internal coating:** The internal coating of epoxy painting confirming to International Standardization Organization (ISO) 15741 (applicable standard) is applied. The minimum dry film thickness must be of about 80 microns.

### *3.4.1.6 Pig Launcher and Receiver*

It must be of appropriate standard (e.g., American National Standards Institute (ANSI) class 600). Higher classes like 900 also may be used according to the design for operational requirements. In case of reverse flow, it must be able to work interchangeably (i.e., launcher as receiver and receiver as launcher). It will have quick opening end closure, pig signaller, and so on.

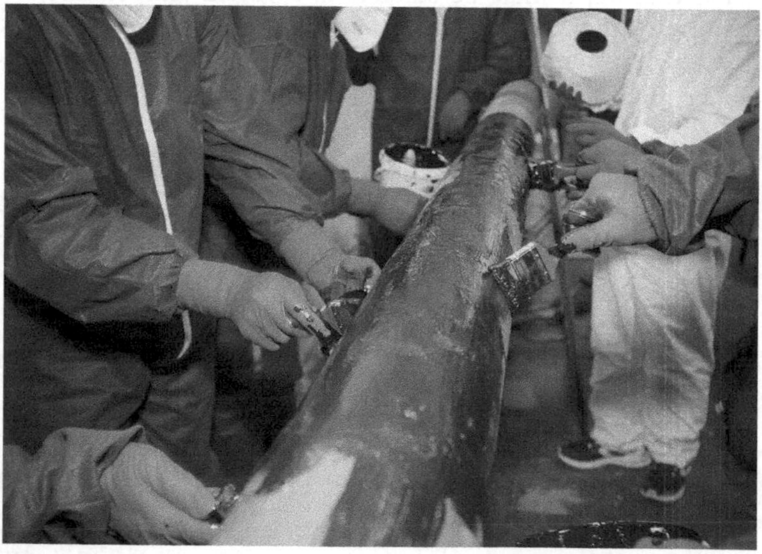

**FIGURE 3.7**
Joint coating activity of 3 LLP coated 24″ diameter pipe. (From www.gspcgroup.com.)

#### *3.4.1.7 Flow Tees*

All branch connections at more than 40 percent of the diameter of the main line diameter must be with a flow tee. The design of flow tee must be according to the American Society of Mechanical Engineers (ASME) standard, which should allow smooth passage of pigs. The flow tee material must be compatible with the pipe material.

#### *3.4.1.8 Piping*

Piping class must be in accordance with ANSI class 600/300/150 and according to piping and instrumentation diagrams (P&ID).

#### *3.4.1.9 Insulating Joints*

Insulation points are provided at transition points of above ground and underground portions of the pipeline for electrical isolation. It must be of monolithic block type.

#### *3.4.1.10 Sectionalizing Valves and Ball Valves*

SVs and scraper trap isolation valves must confirm to API 6D and be a ball valve of full-bore type for smooth passage of scraper pigs. The SVs must be buried and provided with gas actuators. All other SVs must be operated locally.

Ball valves must confirm to API 6D and be of full-bore type for smooth passage of scraper pigs. They are underground for SV stations and above ground for IP stations.

SVs and ball valves may be 30-, 24-, or 18-inch nominal bore (NB) sizes depending on the diameter of the main trunk line and of the 600 #/900 # rating. It is for IP stations, dispatch stations, receiving stations, and SV stations.

### 3.4.2 Civil and Structural Work

The civil and structural work is undertaken according to the prescribed and applicable standards and codes. The following civil/structural work is required:

1. Buildings
2. Process Area
   a. Valve operating platforms and crossovers
   b. Pipe and valve support structures
   c. Equipment foundation

3. General civil works like cable trenches, reinforced cement concrete (RCC) pavements, roads, storm water drains, boundary walls, and so on
4. Miscellaneous works such as breaking and improving the existing roads, RCC works, and so on, whenever pipe/electrical/instrumentation cable crossing are required

The land along the pipeline-laying route needs to be flat and cultivated or free of any permanent structures. All major river crossings are managed by horizontal directional drilling (HDD; trenchless construction method) method or concrete coated pipes; national highways and railway crossings are considered as cased crossings or HDD. There may also be chances of crossing another foreign pipeline. The methodology of crossings is to be followed according to the permissions and agreements with the authorities concerned. Pipeline-laying costs may include the cost of laying of optical fiber cable (OFC) and high-density polyethylene (HDPE) conduit. The cost for erecting station equipment and piping is included in pipe-laying cost. While designing all the previously mentioned structures, the life of the structure is considered according to the regulatory guidelines (Figures 3.8 and 3.9).

**FIGURE 3.8**
Transmission line. (From Turkish Petroleum Pipeline Corporation—BOTAS, Ankara, Turkey.)

**FIGURE 3.9**
Fabrication of station piping installation of mainline valve and bypass line.

### 3.4.3 Electrical Work

The electrical work consists of the following which is carried out as required according to applicable standards.

1. Power supply source for essential loads and non-essential loads
2. Earthing system
3. Lighting system
4. Cables
5. Air conditioning of control room

While planning, the essential loads are considered for the cathodic protection system, telecommunication system, instrumentation and control, SCADA, remote terminal unit (RTU)/PLC, fire detection and alarm system, and auxiliary power for the compressor. Thermo Electric Generator (TEG) and close circuit television (CCTV) power also is considered for essential process loads at IP stations and at all remote operated SV stations.

Non-essential load is considered for lights and fans, the air conditioning system, and the power socket for maintenance. The air-conditioning system as such is an essential part because it prevents the systems from heating up in the control room. Grid power is considered for non-essential loads. UPS and diesel generator sets are planned for manned stations.

### 3.4.4 Instrumentation Work

The instrumentation system is designed considering the site conditions and applicable standards. The system consists of field instruments and a fire and gas detection system. The work is related to:

1. Engineering, supply, erection, and commissioning of field instruments and the fire alarm system
2. Cable laying through cable tray and cable trench from field to PLC-based control panel through junction boxes and for inter-panel wiring inside the control room with proper gladding, termination, ferruling, and so on
3. Installation of junction boxes
4. Calibration of all instruments, leak test, and so on
5. The fire detection and alarm system

Field instruments are field transmitters, pressure gauges, resistance temperature detectors (RTDs), control valves, Current to Pressure (I/P) converters, metering system, gas chromatograph, instrument cables, and (MCT), blind PLC panel, and PLC-based control desk.

### 3.4.5 Supervisory Control and Data Acquisition System (Control Room)

The pipeline is monitored and controlled from the SCADA system which consists of a master control station (MS), a secondary master control station (SMS), and a remote work station (RWS). It ensures effective and reliable control and management and supervision of the pipeline. The work relates to (1) engineering, supply, erection, and commissioning of the SCADA system and (2) interfacing of all instrumentation signals with SCADA through RTU/PLC through a serial link and a hard-wired link according to P&ID.

All stations have RTU for field signal connectivity with SCADA. The serial link is provided for flow computer (FC), gas chromatograph, and CP system signals. However, flow rate, line pressure and temperature (for FC) and pipe-to-soil protection (PSP) value, impressed voltage, and current (for CP) is hard wired.

The following applications (APPS) modules and functionality can also be provided.

1. Leak detection and leak location
2. Inventory analysis
3. Survival time analysis
4. Pipeline transportation efficiency and scraper tracking modules
5. Contingency analysis

6. Planning module
7. Predictive module
8. Compressor fuel optimization
9. On line network simulation
10. Flow management system

One dedicated server and one man machine interface (MMI) is installed at the Master Control Station for the APPS system. The APPS system is designed independent of the SCADA system hardware and software and provided with the necessary tools to access the real-time database of SCADA to fetch the required data for APPS. The networking needs to support open systems interconnection/transmission control protocol/internet protocol (OSI/TCP/IP) network connectivity (Figure 3.10).

### 3.4.6 Telecommunication System

The telecommunication system work consists of the following:

1. Engineering, supply, erection, and commissioning of the telecommunication system along with all necessary accessories
2. Optical fiber cable (OFC) termination for the telecommunication system
3. Interfacing of all instrumentation signals with SCADA through RTU as per P&ID

**FIGURE 3.10**
Central control room/master control station. (From www.gspcgroup.com.)

The telecommunication system must provide voice, data, and video communication. The system must also have optical fiber cable, optical transmission equipment, primary/drop–insert multiplex equipment, network management system, private automatic branch exchange (PABX), telephone instruments, digital telephone, weather-proof telephone, explosion proof telephone for hazardous area and acoustic booths, a video conferencing system, and so on.

### 3.4.7 Cathodic Protection System

The electrical equipment and systems used for cathodic protection system are (1) power source with battery back-up, transformer rectifier unit, or thermo electric generators, (2) anodes for temporary cathodic protection (magnesium ribbon anodes), (3) mixed metal oxide anodes used for impressed current cathodic protection (ICCP) for permanent cathodic protection (PCP), (4) cables, (5) polarization cell, (6) half-cell reference copper sulphate electrodes, (7) test lead points, (8) insulation mono blocks, and (9) polarization coupons.

## 3.5 Project Financial Analysis

### 3.5.1 Project Cost Estimates

The break-up of aggregate cost estimates (as percent of total cost) is given in Table 3.1.

Approximate cost for natural gas transmission pipeline projects may be about USD 5 to 7 million per line kilometer (as of the year 2017/18). It may vary from country to country. However, it is mentioned here to get some insights about the project cost at the time of initial and planning phase of the project.

To work out the cost of land, the area of land required may be worked out as follows:

1. SV stations may be of 40M × 30M each (manually operated, remote operated, and tap-off locations)
2. IP stations of 65M × 65M, SV stations with a compressor of 300M × 250 M
3. Dispatch stations of 85M × 80M and receiving station of 85M × 70M
4. Stations may also be provided with area of 30M × 40M for a helipad

**TABLE 3.1**

Break-up of Aggregate Project Cost (% cost of total cost)

| Sr. No. | Description | Aggregate Cost (%) |
|---|---|---|
| 1 | Pipeline Route Survey and Soil Investigation | 00.09 |
| 2 | RoU and Crop Compensation (this is usually based on 30 M width along the pipeline) | 00.60 |
| 3 | Line Pipe with Internal and External Coating | 37.00 |
| 4 | Ball Valves and Sectionalizing Valves | 00.97 |
| 5 | Line Material (Flow Tee, Insulating Joints) | 00.09 |
| 6 | Pipe Laying and Station Works | 18.00 |
| 7 | Cathodic Protection System | 00.30 |
| 8 | SCADA/Telecommunication System including OFC | 01.25 |
| 9 | TEG/CCTV | 00.42 |
| 10 | Filtration and Metering System (installed at dispatch terminal and receiving station) | 00.50 |
| 11 | Compressor | 03.50 |
| 12 | Scraper Traps and Pig Signaller | 00.20 |
| 13 | Piping | 01.25 |
| 13.1 | Electrical and UPS | 00.35 |
| 13.2 | Instrumentation | 00.25 |
| 13.3 | Spares | 00.25 |
| 14 | Civil Works | 00.80 |
| | Freight, Insurance, Taxes, and Duties (Indirect Cost) | 15.58 (May be added in above all proportionately) |
| 15 | Project Management (Detailed Engineering, Procurement, and Construction Supervision) | 02.50 |
| 16 | Owner's Management Expenses at 1.5% of total project cost | 01.20 |
| 17 | Start Up and Commissioning Lump Sum | 00.60 |
| 18 | Line Pack (It is a procedure for allowing more gas to enter a pipeline than is being withdrawn, thus increasing the pressure, "packing" more gas into the system, and effectively creating storage) | 01.00 |
| 19 | Contingency at 3% on cost of civil works and 5% on the cost of other items | 05.00 |
| 20 | Working Capital Margin | 00.30 |
| 21 | Price Escalation | 05.00 |
| 22 | Interest During Construction | 03.00 |
| 23 | Financing Charges of lending agencies | As per prevailing rates |
| 24 | Township if planned | Actual cost as per planning |

### 3.5.2 Project Economic and Financial Feasibility

To assess the economic feasibility of the project the following parameters are analyzed for the period of economic life of the project (i.e., 25–40 years):

1. Demand for natural gas (mmscmd) (yearly) in (a) the first tariff zone, (b) the second tariff zone, and (c) the third tariff zone up to 25–40 years
2. Volume of natural gas to be transported in mmscmd (yearly) in (a) the first tariff zone, (b) the second tariff zone, and (c) the third tariff zone (yearly)
3. System capacity of pipeline as per relevant regulation (mmscmd)
4. Natural gas pipeline tariff (yearly): (a) natural gas pipeline tariff (Rs/Million Thermal British Unit) in first tariff zone, (b) percentage increase in tariff for second tariff zone over the first tariff zone, and (c) percentage increase in tariff for third tariff zone over the second tariff zone
5. Total cash outflow (USD): (a) CAPEX, (b) operating expense (OPEX)
6. Total cash inflows (USD): natural gas pipeline tariff
7. Pre-tax internal rate of return (IRR) on total capital employed based on the previously listed project cash flow

We need to note that variables (2) and (4) as shown are biddable.

The assumptions which may be considered for the financial analysis are:

1. The total life of the project may be considered as 25–40 years from the date of commencement of supply of gas.
2. Project gestation period may be taken as 2 years from the project approval date.
3. Terminal value of 5 percent may be considered at the end of the life of the project.
4. The demand for gas at certain level as expected. The variation in this has an impact on IRR of the project.
5. The debt to equity (D/E) ratio (may be considered as 60:40 or according to the regulations in respective country), interest rate, moratorium, inflation rate, and repayment term for the debt may be assumed appropriately for financial projections.
6. The assumptions for operating and maintenance (O & M) cost may be considered as (a) 1 percent of (line pipe material) + 3 percent of balance equipment, (b) the manpower for control stations and compressor stations, (c) appropriate fuel cost, and (d) escalation cost.
7. The depreciation at the appropriate rate may be considered. It must be according to government regulations.

8. The insurance cost at the rate of 1 percent per annum on a replacement basis may be considered for a profit and loss (P&L) account statement. An annual inflation may be considered appropriately.
9. The selling and marketing expenses may be considered as 1 percent of annual revenue.
10. The interest on working capital may be considered appropriately and 30 percent of working capital may be capitalized in project cost as a working capital margin.
11. The corporate income tax may be considered in line with the country's income tax structure.

### 3.5.3 Methodology for Fixing the Tariffs

To determine the base tariff and variable tariff following methodology is used:

1. The methodology is based on requirement of the business to achieve the minimum IRR based on fixed and variable tariffs. The calculations may be carried out based on a half yearly or monthly basis.
2. To achieve a fixed tariff all expenses, namely, (1) O&M expenses, (2) Depreciation, (3) insurance, (4) payment of loan interest, and (5) selling and marketing expenses are worked out. Thereafter, based on flow of gas, a tariff is estimated, and IRR worked out. The tariff which gives the desired IRR is considered the base tariff.
3. To find variable tariff, the methodology is same except estimated ROE is considered instead of estimated tariff and when minimum IRR is achieved the tariff is calculated based on the revenue generated divided by the flow of gas. The average of all the variable tariffs is considered the base tariff.

### 3.5.4 Sensitivity Analysis

The sensitivity analysis is carried out based on (1) cost overrun and (2) reduction in gas volume.

The analysis will help the project sponsor check the viability of the project and useful for bidding.

## 3.6 Project Implementation Strategy

The pipeline projects usually run into hundreds to thousands of kilometers. The implementation of such large projects within a specified time schedule and budget requires a well-defined strategy to implement the project effectively.

It requires effective coordination and communication between the stakeholders and planning and developing suitable monitoring and control mechanism for all engineering, procurement (special attention may be given to long lead items), manufacturing, supply, erection, and commissioning activities of all disciplines (i.e., civil, mechanical, electrical; instrumentation and control [I & C]). Usually such projects require the services of a specialist-engineering consultant and a project management consultant.

In projects of such magnitude, the implementation, construction, and erection will require execution of multiple construction spreads by more than one contractor. Each construction spread will need dedicated construction manager. The teams of each spread need to coordinate with the general construction manager (overall in-charge of the project) who coordinates, manages, plans, and monitors the overall project progress. This coordination helps assure homogeneity in construction management, quality, and so on in all spreads.

Development of detailed planning schedules at various levels of activities is required during the implementation stage using project management software (i.e., primavera or MS project). The software will show linkages between various activities. Dedicated teams of planning engineers will track and monitor progress and implementation of these schedules. The teams inform the appropriate levels of management about the progress achieved versus targets, any variations, and criticalities through daily, weekly, and monthly reporting for early resolution of any delays.

### 3.6.1 Engineering Management

The project engineering (i.e., basic engineering and detailed engineering) is done before implementation of the project.

#### *3.6.1.1 Basic Engineering*

The conceptual engineering design and basis of design is a key document for proceeding with the detailed engineering of the project. The document includes pipeline system structure, the route of the proposed pipeline, and location of stations and metering system, SCADA, and telecommunication system.

The soil testing report for the proposed route is the key input data to commence the basic engineering of the project.

After the project is approved, the conceptual engineering design and basis of design established during DFR is reviewed by the engineering consultant. In case of any change in base parameters, they are frozen in consultation with the project sponsor.

The conceptual engineering basically establishes the process design basis. Based on the system analysis, the various process flow diagrams (PFDs) and the equipment list are elaborated.

Design basis for various disciplines (i.e., civil, electrical, instrumentation, SCADA, and telecommunication along with piping and instruments diagrams [P & IDs]), basic layout of stations, and electrical single line diagram (which is the base document for further engineering) needs to be finalized by priority by the engineering consultant. To finalize the basic layout of stations, the plots for various terminals and stations needs to be identified.

Identification and procurement of the long lead items (which may be the critical equipment and require a long period to manufacture and ship) need to be initiated at a very early stage of engineering. Therefore, engineering for these items may be completed on a priority basis.

The basic engineering is proposed based on the packages. The packages may be specified by the functional requirement, duty specifications, protections, and so on. The data sheets and a general specification covering the general terms and condition and the general technical requirement of packages are also specified.

#### *3.6.1.2 Detailed Engineering*

The detailed engineering proceeds after basic engineering. The basic engineering concepts are developed into greater details through detailed engineering for construction of the project.

A procedure needs to be developed for effective engineering coordination for development of all drawings, documents, and correspondence related to project engineering in a uniform manner. This procedure promotes understanding and interpretation uniformly by all the concern parties.

### 3.6.2 Procurement Management

Procurement is done in terms of various packages. The main packages identified are given in the following list. The packages need to be further detailed by the engineering and project management consultant.

1. Main coated pipes
2. Mainline ball valves
3. Other station valves
4. HDPE duct
5. Optical fiber cable
6. Station pipes
7. Insulation joints
8. Pipe fittings
9. Metering and filtering skids
10. Scraper trap

#### *3.6.2.1 Work Packages*

Various contractors appointed for different activities undertake the construction at the site. The following major works packages may be identified and need further review before project implementation which also includes some critical and specialist turnkey jobs, which may involve the entire engineering, procurement, and construction (EPC) by identified contractors:

1. Mainline and station construction works, mechanical and civil
2. Electrical and instrumentation works
3. Triethylene glycol (TEG) unit/closed circuit television video (CCTV) works
4. Cathodic protection works
5. SCADA and telecommunication works
6. Gas turbine (GT) driven compressor works

The intermediate compressor stations may be taken up at later date.

#### *3.6.2.2 Priority Procurement Items*

The following items which are required to start the project work are priority procurements:

1. Coated line pipes
2. Optical fiber cable
3. HDPE duct
4. Works package for mainline and station works
5. Cathodic protection works

Long lead items are also identified and addressed in the beginning to avoid delivery bottlenecks. Following could be the long lead items to address on priority.

1. Mainline ball valves
2. GT-driven compressor works packages
3. Insulation joints
4. Mainline and station pipes
5. Scraper traps
6. Metering and filtering skids
7. Bends and so on

#### *3.6.2.3 Procurement Procedure*

Development of a detailed procedure is required for the procurement and for expediting the project work. The procurement guideline may include the following:

1. Preparation of bid documents (technical and commercial)
2. Preparation of cost estimates
3. Notice inviting tender (NIT) as per firm's guideline
4. Opening of the bid documents
5. Technical and commercial evaluation (may have intermediate stage of queries and clarifications which can be managed through pre-bid conference and planned before submission of bid by the parties)
6. Bid recommendation
7. Award recommendation
8. Letter of award
9. Agreement

#### *3.6.2.4 Expediting Procedure*

After the order is placed, the expediting activities commence to supply equipment (covered under procurement packages) to the construction site. The activities must match with the site requirements and the project schedule. The expediting team also needs to ensure that the inspection and certification is carried out according to the finalized quality assurance plans.

#### *3.6.2.5 Quality Control and Inspection*

The quality of various products must be based on the approved quality assurance plan. A base and minimum quality assurance plan may be included in the various procurement packages at the tendering stage. The final quality plans for inspection by the vendor at the vendor's premises or at the vendor's designated sub-supplier will need to be finalized with the vendors after award of the contract in line with the base quality assurance plan (QAP).

To meet the requirement of the items over the project life cycle, every item must be supplied according to the tender specification which meets the required strength with the desired properties and performs operational requirements over the full life of the project. The items must also work under fluctuating process requirement and climatic conditions.

Detailed quality plans and procedures need to be developed separately for imported items and for the indigenous items.

### 3.6.3 Construction Management

Gas transmission facilities need to be constructed such that the pipeline meets the required quality level and the necessary reliability in the long run and does not present undue risks for the people and the environment. All pipeline and systems are designed, constructed, and tested according to ASME standards. Welding may be carried out according to API standards. To expedite the construction of mainline stations, automatic welding is recommended.

Apart from the construction contracts, construction management plays a major role in the management of the quality control and day-to-day supervision of the construction activities. The project management consultant appointed by the project sponsor has a key role in construction management. A suitable construction management organization needs to be developed.

The following construction methodology may be used for implementation of a project.

#### *3.6.3.1 Construction Spreads*

The first step is to plan the number of spreads based on the following considerations:

1. Overall time frames for the project
2. Construction of other similar projects of sponsor and logistics requirements
3. Geographic and inter-state layout of the pipeline. The spread alignment with interstate boundary is important considering the tax regime

As mentioned previously, the construction of a long pipeline requires deployment of more than one contractor.

#### *3.6.3.2 Construction Supervision*

The construction supervision must be carried out by the Project Management Consultant (PMC) who the project sponsor appoints. The PMC usually identifies a general construction manager before implementation. The general construction manager will be in charge of all the construction activities of the project. The role of the general construction manager is (1) to achieve uniform norms of construction and quality across all spreads, (2) to achieve judicious transfer, use, and control of resources across all spreads, and (3) to bring innovative construction concepts to the project.

The PMC will ensure that the construction manager of each contractor who manages the individual spread is responsible to general construction manager. Further, each of the spreads is managed by a spread in-charge.

A team of mechanical, civil, electrical, instrumentation, planning, and non-destructive testing (NDT) engineers assists each spread in-charge. The health, safety, and environment (HSE) norms at each spread is the responsibility of the HSE manager who will be assisted by safety officers. Each spread will also have its stores managed by a store supervisor assisted by a required number of stores assistants. All the discipline engineers, HSE, and stores are the responsibility of the spread-in-charge.

A surveyor is required for the initial construction period to carry out route survey, finalize route alignment, establish area requiring special attention during construction, provide information on soil, requirement of buoyancy measures, site bends, and locations of TCP/IP and so on, RoW, and the nature of construction along the route.

The requirement for supervision of various construction activities generally can be categorized as:

1. Those activities which must be monitored critically because of the direct impact on the integrity of the pipeline in the short and long term. These may be categorized as critical construction activities and others.
2. Activities which can be audited during the works and checked after the work finishes by using batch checks.

The critical activities and the batch activities for the construction supervision must be finalized at a later stage during the project implementation.

#### 3.6.3.3 Quality, Health, Safety, and Environment

For the gas pipeline project which is of very large scale, quality, health, safety, and environment (QHSE), concerns are very important during the project execution stage. The contractor needs to adhere to the QHSE guidelines. QHSE guidelines usually are developed and included in the tender. Based on these guidelines, the contractor will submit a relevant project quality plan and HSE procedure for approval from the project sponsor and PMC.

The objective of the project quality plan is to ensure that all phases of the project are executed with the uniform level of quality and efficiency which ultimately will satisfy the project sponsor's requirements. The quality plan must also comply with all relevant and applicable regulatory standards and codes. The completed facilities and the related equipments are handed over to the project sponsors as a safe, reliable, and efficient operating facility. The project sponsor or owner and the contractor for the same will agree upon a suitable check list for the plan.

The HSE plan or procedure covers the identification of various hazards associated with various pipeline activities and measures to avoid the occurrence of hazards. It also includes details of preparation required to counter the hazard in case it occurs and the measures to ensure the health of the various persons involved in the project. Due consideration is also given to establish the impact of construction activities on the environment and measures to mitigate or minimize its impact.

### *3.6.3.4 Stores Management*

The stores for construction management must be located centrally as close to the work site and as centrally to the spread as possible. It must have road accessibility. All necessary facilities such as stores management software tools and other utilities are provided.

Storage space is required at each store to stock station piping material and large items such as metering skids. Adequate security arrangements must be provided in the stores during the construction stage. These must remain in force until the closure of the project. The development of the stores management procedure addresses the procedure for material received at the main stores and the distribution of the material to the contractors.

Following are the challenges faced by the firms while planning natural gas transmission pipeline networks:

1. Optimization of pipeline diameter and layout and route which is based on demand over the next 10, 15, 25, or 40 years. Cost optimization must be done by choosing higher diameter pipeline or smaller diameter pipeline with compressors at boosting stations. Furthermore, pipelines with various diameters are chosen based on the clustered demand of different markets. Selection of pipeline layout also plays a very important role through proper planning because not only is capital cost saved but also operations and maintenance cost is saved. Layout must be selected considering gas hydraulics as well.
2. Operating pressure plays a very important role as this determines the pipeline thickness and number of compressions required. Meeting the demand at lower operating pressure provides substantial saving for the overall project.
3. Positioning of various equipment is a challenging task because it increases the efficiency of pipeline. Positioning of compressors optimizes the pipeline throughput. Positioning of other equipment like valves and regulators at strategic location optimizes the maintenance cost, apart from maintaining the pipeline throughput.
4. Quality of materials is a challenge for planning a pipeline transmission network. Therefore, pipeline, valves, compressors, regulators, and so on are chosen based on the overall quality for the entire lifecycle of the project rather than only on cost.
5. Contractors for the pipeline network are at the forefront in execution of pipeline transmission network; therefore, selecting a contractor

who can complete the project on time without any cost overrun is of prime importance.

6. Identifying and deploying experienced manpower in a cost optimized manner is also a challenging task.
7. Permission for RoW is a very critical task and the success of the whole project depends on getting land permissions from many agencies and farmers. Pipeline can be laid only after getting permission for land use.
8. Obtaining statutory clearances for pipelines crossing rivers, canal, railway crossings, and so on from respective departments and authorities on time is critical for timely completion of project. Compliance with environmental clearance is also one of the critical requirements for starting the project.
9. Issues raised by the local people must be properly addressed such as compensation for their land and crops according to the prevailing regulations. Some of the Corporate Social Responsibility activities in the localities help build trust between the company and the local population.
10. Gas pricing play a very important aspect as it decides the usage by the end customers as fuel. The uncertainties in gas pricing are a big challenge for pipeline firms as the volume to be transported for the users of the pipeline depends upon the Gas price. The transmission business earns the tariff for volume of gas transported.

## Bibliography

De Wolf, D., and Smeers, Y. (1996). Optimal dimensioning of pipe networks with application to gas transmission networks. *Operations Research*, 44. doi:10.1287/opre.44.4.596.

Gunes, E. F. (2013). Optimal design of a gas transmission network: A case study of the Turkish natural gas pipeline network system, Graduate thesis & dissertation, Iowa State University, Ames, IA.

Jain, C. (2007). Internship project report on gas pipelines optimisation, Institute of Petroleum Management Gandhinagar, Gujarat, India.

Martin, A., M. Moller, and S. Moritz. (2006). Mixed integer models for the stationary case of gas network optimization. *Mathematical Programming*, 105, 563–582. doi:10.1007/s10107-005-0665-5.

Rothfarb, B., H. Frank, D. M. Rosenbaum, K. Steiglitz, and D. J. Kleitman. (1970). Optimal design of offshore natural gas pipeline systems. *Operations Research*, 18(6), 992–1020.

www.gspcgroup.com (accessed on August 18, 2018).

www.gailonline.com (accessed on August 25, 2018).

# 4

# *Natural Gas Transmission Business: Operations and Maintenance Aspects*

This chapter presents the basic philosophy and key criteria for operation of the natural gas transmission pipeline and associated equipment. Some basic guidelines to operate, maintain, and control the transmission pipeline system are discussed. The firms in this business must develop detailed operating procedure for operations, maintenance, and control. The entire gas transmission pipeline system and related facilities are to be planned such that it is under supervision and monitoring of a central control room (CCR).

## 4.1 Operations of Main Equipments

### 4.1.1 Filter

Filters are located at compressor stations and metering stations. Two or more filters usually are used to allow maintenance of the maximum designed throughput so that if one filter fails then the other one (i.e., standby) takes over the function of the failed one.

A specific system is provided in the control system for the filter units. The system enables the operator to select the required filter train automatically and default to stand-by. The differential pressure across the filter train is monitored by the local control center and, if required, initiates an alarm indicating the necessity to switch to the stand-by filter upon reaching high differential pressure. The operator can make a judgment call to initiate the switching of filter trains. Certain types of filters have a coalesce element and therefore have small liquid boot. If this type of filter is used, it will require level indication and alarms for high liquid level build-up (Figure 4.1).

**FIGURE 4.1**
Filters. (From www.balstonfilters.com.)

### 4.1.2 Meters

Metering facilities are located at the inlet terminal, at intermediate pigging stations, at end user stations with pressure reduction stations (i.e., high-integrity pressure protection system (HIPPS) and/or flow control) and compressor stations. The meter skid facilities operate as an independent unit which is linked to the control system through its own panel. The measurement is based on ultrasonic technology (Figure 4.2).

**FIGURE 4.2**
Meter and custody transfer skid. (From www.controlplus.in.)

### 4.1.3 Compressor and Compressor Control System and Natural Gas Turbine

The operations of compressors and other associated equipments are performed by the compressor control system. The compressor control system consists of Unit Control Panel (UCP) for each compressor and a proprietary load-sharing controller.

The compressor control system will perform the functions of an individual compressor load, anti-surge control, and overall control between compressors.

The process control system interfaces with the compressor control system through dual redundant communication links for the transfer of status and alarm information. All the commands from the station control or integrated safety system to the compressor control system are in the form of discrete hard-wired signals.

The compressor control unit is a part of the compressor and turbine package. All control functions associated with the compressor and turbine are managed through the compressor control unit.

The compressor control unit can start and stop the compressor and turbines, monitor performance, and help in safe guarding the equipment. It also helps in automatic switching to any stand-by compressor. The compressor control system exchanges key data to and from the supervisory control and data acquisition (SCADA) system.

The compressor control unit further ensures that the suction pressure is maintained during start-up and normal running. It also monitors and controls the discharge pressure of each compressor. Flow rate is also monitored along with pressure differential across the compressor for surge control.

The compressor train is controlled for flow requirements and pressure limitations (i.e., it attempts to meet the flow demands), but the pressure control will take over if the pressure limits are reached.

Large machines such as compressors with their drivers are condition monitored. Alarms from these systems and status indications are sent over the SCADA system if the automatic trip is initiated. However, it is a normal practice to download the remote off-line diagnostic information (not in real time) separately over the communication links, independent of SCADA system. This information is then analyzed by machinery specialists to predict the time for the next maintenance, whether it is safe to restart a machine after a trip, and so on. This practice saves unnecessary maintenance and minimizes unscheduled outages, which can save substantial production losses.

Natural gas turbines are used to generate power using natural gas as a fuel. The captive power is generated to operate the compressors and other pipeline equipments. The capacity of the turbine is planned in line with the power requirement to run the transmission line system with all connected equipment. The turbine will have all the required auxiliaries as that of a power generation unit.

### 4.1.4 After-Cooler Control System

The after-cooler control system for each compressor train monitors the outlet-temperature of the gas and switches the cooling fan motors on or off based on the required set-point temperature (which could be 50 degree Celsius). A suitable bandwidth is set to prevent continuous starting and stopping of the fan motors. An alarm is provided for high temperature (60 degree Celsius) and for very high temperature (70 degree Celsius) wherein shutdown is provided to protect the pipeline network. In case of shutdown the system isolates the associated compressor train.

### 4.1.5 Pigging Operations

Pigging operations are carried out for the following activities:

1. Post installation pigging operations are carried out for hydrostatic testing to fill the pipelines with test water and subsequently remove the water.
2. During operations pigging activities are carried out to remove dust or debris in the pipeline. The frequency of pigging depends upon quantity of debris found during operations.
3. At about 5-year intervals to check the pipeline condition using an intelligent pig. Intelligent pigs have sensors attached to them which collect data about metal erosion, thickness of pipeline, any damages to pipeline, and so on. This information helps for taking corrective actions.

The main transmission pipelines and laterals or branches (depending on diameter and length) have either fixed or removable pig traps installed, along with associated pipeline network. Pigs are normally made of rubber with Teflon coating, core metal with bushes, and magnets. Standard operating processes (SOPs) are developed to ensure efficient and safe operations for pigging (Figures 4.3 and 4.4).

### 4.1.6 Fire Fighting System

A fire water system is provided at the compressor station. Automatic fire detection equipment triggers alarm systems in the event of a fire. The fire water system gets activated to provide a fire-fighting capability. The fire water system for the compressor station includes a main firewater loop, hydrants, and monitors.

The enclosure of the compressor is protected with automatic $CO_2$ fire extinguisher systems. To provide required fire protection inside the turbine, compressor, and generator building, $CO_2$ and dry chemical extinguishers are installed.

**FIGURE 4.3**
Pig. (From www.naturalgas.org.)

**FIGURE 4.4**
Pig. (From www.pgjonline.com.)

### 4.1.7 Pressure Reduction Station

The pressure reduction stations (i.e., HIPPS and flow control) are required for the end user facilities. The stations are manned (by guard) and are controlled locally and automatically through the SCADA system. The system is comprised of two or more filters which are designed such that maximum throughput can be maintained with one filter in case of failure of one of the filters. Failure of one filter automatically switches to the stand-by filter. The filters may be designed for the maximum inlet pressure and maximum differential pressure at the coldest natural gas arrival temperature.

The gas filters used may be water shed and tube style. The by-pass temperature control valve controls the flow of water through the filter. Each valve receives signals from a temperature controller downstream of the discharge point from the filter and a temperature controller downstream of the pressure reducing station.

Pressure regulators with associated slam shut valves may have a two in series and two in parallel arrangement. Each stream will have a slam shut (1st stage) monitor valve and an (2nd stage) active valve. If failure occurs, then the second train will be brought automatically on-line to replace the failed one.

Metering may have spare trains that can be brought automatically on-line in case of failure of one of the operating streams. The pressure regulation and metering consist of enough streams to handle the initial and ultimate flow rate at each phasing (Figure 4.5).

### 4.1.8 Emergency Shutdown System

The function of emergency shutdown system (ESD) is to protect personnel, equipment, and environment from the consequences of an accident or an uncontrolled release of gas and/or any other process.

FIGURE 4.5
Pressure reduction skid. (From www.controlplus.in.)

ESD facilities are provided at the inlet terminal station and the compressor station. It is the current practice wherein the ESD system is not installed at block valve stations or off-take stations considering the safety and hazard and operability (HAZOP) studies.

The pressure reduction skid at the off-take stations have slam shut valves in the system to protect the downstream pipeline network from excessive pressure if a pressure regulator stream fails.

### 4.1.9 Depressurization System

A depressurization (i.e., venting) system is required for a compressor station to reduce the amount of hydrocarbon in the pipeline system in the event of fire, explosion, or any other dangerous situation that may occur. The purpose of the system is to reduce the pressure to minimize the probability of vessel fracture in case of fire. The gas released at depressurization is vented out to safe location through a common blow-down and pressure relief system to the vent stacks.

A permanent vent stack is suggested also at the block valve stations and/or end user stations. The vent installations are placed at a safe distance to vent these stations and underground sections. The rate of depressurization is controlled considering the wind direction, neighboring conditions, and the kind of intervention (emergency, maintenance, commissioning, and start-up) required.

### 4.1.10 Valves

Large ball valves are opened when the differential pressure across them is less than 5 bars. This pressure must be capable of being reset between 1 and 5 bars. The control system inhibits the remote opening of several of the valves until the pressure is balanced. This control is to prevent erosion of the valves seats and avoid pressure shock in the downstream network.

Pressuring by-pass is provided around all valves that could be opened with a differential across them. The by-pass consists of an upstream and downstream ball valve, a globe valve, and a vent system between the two. The SOP for opening valves is as follows (Figure 4.6):

1. Ensure vent valve on the by-pass is closed
2. Unlock ball valves on the by-pass (if required)
3. Open the ball valves
4. Slowly open the globe valve to pressurize downstream section
5. Monitor downstream pressure indicator to determine when the pressure is balanced
6. Once pressure is balanced, open the large bore valve

**FIGURE 4.6**
Ball valve. (From www.Amtechvalves.com.)

## 4.2 Maintenance of Natural Gas Transmission Lines

Pipeline and station maintenance (i.e., inlet terminal, block valves, end user stations, and compressor stations) is an important element to ensure that the project infrastructure and pipeline gas grid operate at maximum efficiency and minimum cost with minimum downtime. In addition, the implementation of a maintenance program ensures safe operation and continuity of gas supply to the customers.

The maintenance program must be recorded using the software and computers which may include inventory, spares, and components. The program also must maintain the record of maintenance schedule with details of maintenance and a record of malfunction of the transmission line system.

The major activities which must be undertaken for pipeline and stations maintenance are discussed in the following section.

### 4.2.1 Pipeline Maintenance

The pipeline is inspected regularly, and the maintenance system is designed to cover the main elements through visual surveys, security and safety surveys, coating defect surveys, and intelligent pig surveys.

#### *4.2.1.1 Visual Surveys*

Visual inspection survey of the pipeline is an important source of information for pipeline maintenance. The following aspects may be reported at an appropriate interval during daily patrols along the pipeline right of way (RoW).

1. Any signs of security or safety problems
2. Any signs of subsidence at road or rail crossings
3. Soil erosion or removal of soil cover from the pipeline
4. River or stream bed changes which reduce the cover to the pipeline
5. Construction activities close to the pipeline network
6. Increased population near the pipeline network which may impact the design factor

Apart from visual survey, aerial survey of the pipeline is carried out also to identify any unusual activities near the pipeline and, if required, further inspection on ground can be carried out.

The regular liaison with the local authorities is required to ensure that the modification in infrastructure does not impact the pipeline integrity.

The information collected from the inspection surveys must be complied into permanent records for the pipeline system and these records are checked periodically to ensure that recurrence of faults can be regularly detected. The use of computer-generated recording system using geographic information system (GIS) is planned to facilitate entry and retrieval of information.

#### 4.2.1.2 Coating Defect Survey

Survey of pipeline anti-corrosion coatings and evaluation of the condition of pipeline is important to ensure the pipeline integrity. Coating defect survey is carried out at about 5-year intervals to ensure an acceptable coating condition. A direct current voltage gradient (DCVG) survey is carried out above the ground to locate any areas of damaged coating.

#### 4.2.1.3 Intelligent Pig Survey

Intelligent pigs fall into several categories depending upon the type of inspection as in the following list:

1. Caliper pigs: These are the pigs which monitor and record the dimensional changes in the pipeline, such as dents, wrinkles, ovalization, fittings, welds, and valves.
2. Geometry pigs: These are like the caliper pigs except that the pig can monitor the location of the pipe and hence the displacement. This ability is especially useful in areas of subsistence, landslides, and earthquakes.
3. Metal loss inspection pigs: These pigs are the most sophisticated of all the intelligent pigs and are used to detect wall thinning and

cracks due to corrosion and overall wall thickness. Some are capable of recording volume metal losses on both the internal and external surfaces. The tools work on one of the two basic principles, ultrasonic (which only work in liquid lines) or electromagnetism (which works in both liquid and gas lines).

The success of any internal inspection program depends on two factors: (1) the cleanliness of the line prior to running the inspection tools and (2) the skill of operators and engineering personnel entrusted to encoding and interpreting the data. Therefore, if this work is outsourced to any agency, its experience and track record must be scrutinized very carefully.

An intelligent pig usually is run through the pipeline sections at 5-year intervals to detect any areas of corrosion, metal loss, or damage.

A base line survey is carried out shortly after pipeline installation and future survey results are compared to the base line survey to allow any changes to be tracked with time.

### 4.2.2 Cathodic Protection Maintenance

The regular planned maintenance of the system is a must after commissioning, final set-up, and adjustment.

Routine monitoring and maintenance are carried out to ensure that the cathodic protection (CP) systems are working correctly. The task is to ensure that the protection potential is maintained. Most of the monitoring and maintenance activities are undertaken after the commissioning. The operator has the responsibility to maintain it. The CP activities are recommended for routine monitoring, maintenance, and fault detection.

#### *4.2.2.1 Monitoring of Cathodic Protection*

It is necessary to carry out regular monitoring activities to ensure that adequate protection potential is achieved and that the power sources are operating within their temperature limits for the ambient conditions. It is also important that anodes are not being driven at current densities above the design limits as it leads to anode depletion. It is advisable that only qualified CP technician carry out routine monitoring and that the repairs and remedial work are carried out under a qualified supervisor.

Usually the operator decides the frequency of monitoring; however, it is recommended that potential measurements may be taken on monthly basis. Monitoring may include (1) pipe to soil potential (from both permanent and portable electrode), (2) anode current, (3) power source current and voltage outputs, (4) visual inspection of power source unit, (5) adjustment of system output, if required, and (6) effectiveness of isolating joints and continuity bonds.

The CP technician records all monitored data in standard format and then the information is collated and analyzed by the CP engineer. This information is then stored into the database and then the database can be used to monitor the status of the systems. The database helps in planning the maintenance.

Additional monitoring is required if stray current effects develop and direct current (DC) output of the power source increases significantly. Tests are carried out to check that there are no adverse effects. Based on the test report, remedial actions, if required, also are taken.

If significant changes in the pipe potential are noted during maintenance, then the operator is notified to undertake required maintenance activities within the suitable time limit. The faults may include (1) increase in current demand due to coating deterioration, (2) deterioration of anode material, (3) disconnection/damage of cables, (4) incorrect adjustment of the power source DC output, (5) failure of fuses, (6) failure of alternate current (AC) supply, (7) failure of transformer rectifier, (8) reversed connections (after maintenance), and (9) failure of isolation joints, insulated flanges, or bond connections.

### 4.2.3 Station Maintenance

The above ground installations are regularly inspected to investigate any potential problems which could result in downtime or unsafe operation. All records of the inspection program are maintained. Most of the equipment suppliers in their operations manual and instructions state the recommended frequency of inspection which the operator must plan to fit in with other equipment inspection and operation schedules.

The maintenance of the above ground installation may be supported by a specifically organized planning and recording system. Essential data with site identification and item reference number is recorded for every major item of the equipment at every installation. While preparing the maintenance schedule, due account may be made of the following:

1. Any statutory requirement related to a specific item of equipment
2. The latest information on equipment faults and other defects and its frequency
3. Any change in gas quality or flow, which may affect pipeline system
4. Revised maintenance schedules in line with equipment performance and operational history

The listed inspection and maintenance operations include functional and diagnostic checks as well as overhaul of equipment according to the manufacturer's recommendations. Equipment which requires such maintenance are (1) compressor and gas turbines, (2) electric generators, (3) air compressors, (4) valves and actuators, (5) pressure regulators, (6) metering system, (7) gas heaters, (8) filtering equipment, and (9) pig traps.

#### *4.2.3.1 Functional and Diagnostic Checks*

Functional checks are carried out every 6 months wherein either the working or standby streams may be changed over, or the idle stream is tested for effectiveness of operations. All the items of equipment are checked for correct operations. Controllers and control valves are inspected, maintained, and calibrated every 6 months. All safety and isolation valves are routinely operated to ensure that they are not defective and will function when required.

Major station may have diagnostic check every year. The extent of these checks depends upon the operational experience with the specific equipment. The metering system instruments are calibrated monthly and metering turbines may be re-calibrated every 3 years.

#### *4.2.3.2 Minor Equipment Inspection*

All the minor equipments above ground installations may be inspected regularly to check any obvious signs of malfunction of the equipment. The frequency usually is once a week. The factors which affect the inspection frequency are (1) history of breakdown and repair, (2) type of flow system (i.e., single or multiple), and (3) the availability of telemetry to indicate fault conditions.

#### *4.2.3.3 Major Equipment Inspection*

Most of the equipment items may be overhauled completely every 4 years. It is equivalent to every 2 years in the regulating mode (i.e., 2 years standby and 2 years working). If required, the parts are replaced or new units/equipments may be fitted. It is suggested that all elastomer seals be replaced. The suggested program may be considered a minimum requirement and replacement of parts may be required more frequently based on the criticality of the equipment and experience of the operators. Sound engineering judgment must always be used to modify maintenance programs as required.

Maintenance of all components of pig traps, including end closures seals, bleed locks, electrical bonds, locking rings, pig signallers, and fasteners may be carried out every year.

The traps whether temporary or removable may be inspected before use for any mechanical damage due to handling.

All actuators must be maintained and tested at least once every year. Inspection and maintenance of gas over oil operated actuators ensures that the installation is sound.

The maintenance schedule must comply with statutory requirements, if any, for testing and revalidation of pressure vessels associated with valve actuators.

Pressure regulating valves are subject to regular, preventive maintenance inspection and/or test to determine that they (1) offer adequate capacity and reliability of service for the purpose of its use, (2) are in good mechanical condition with no leakages, (3) are adjusted to correct pressure, and (4) are protected against dirt, liquids, freezing, or other influences which affect proper functioning.

All pressure limiting stations and relief devices are subject to systematic periodic inspections and suitable tests or are reviewed to determine that they are in good mechanical condition. Visual inspection is made biweekly.

#### *4.2.3.4 Filter Maintenance*

The filter differential gauge readings are recorded through the SCADA system and locally recorded every week. If the differential pressure approaches the alarm setting, then the next level of maintenance is carried out wherein the filter elements are changed.

Filter elements may be changed any time the slave pointer approaches the alarm setting. It is not advisable to keep them until scheduled maintenance or even clean them and reuse them.

All filter elements are changed annually, and differential pressure gauges fitted to filters are calibrated yearly. The alarms are checked by including a false differential to ensure that the alarm signal is received at the controller counter.

#### *4.2.3.5 Instrumentation and Metering System Maintenance*

The maintenance of the instrumentation and metering system is categorized into two categories: (1) planned preventive maintenance (routine maintenance) and (2) non-planned maintenance.

The planned maintenance follows the predefined schedule of activities which is designed to ensure that the equipment at all sites is fully operational and in functional condition. Non-planned maintenance generally is required in the event of device or system failure. Non-planned maintenance is unpredictable, and an action is required when the device/equipment fails.

The schedule and intensity of maintenance activities vary depending upon the importance of the site. Sites which involve gas purchases into the system and sales from it need calibration of instruments once every 2 months. Wherever sales gas meters are employed the meter accuracy is very important. Turbine meters may be removed from the line and calibrated once every 3 years.

The procedure of maintenance for compressor station equipment may be according to the provision of the manufacturer's instructions and manual for each item of the equipment.

The operator may plan an integrated maintenance system for the main equipment (i.e., gas compressor, gas turbine, electricity generator, and air compressor) and for other utilities.

The maintenance of information system may utilize the latest analytical techniques. The system must help to monitor and trouble shoot the main equipment (gas turbines and compressors) along the pipeline from a central location.

The information system is independent of pipeline control system. It allows the use of special sensors to perform accurate diagnosis. The objective of the system is to help ensure that detailed information is available on-stream for gas turbines, compressors, and associated equipment. It helps analyze the rates of degradation of rotating equipment from established baselines to predict when a given gas turbine or gas compressor may require maintenance to avoid serious problems of failure.

Under normal circumstances, the change in the rotating parts of the compressor can be planned once every 7 years.

Sometimes when the flow conditions change, the rotating parts of the compressors are changed for better efficiency and lower fuel consumption.

## 4.3 Control Philosophy to Manage the Natural Gas Transmission System

The central control room (CCR) assumes the functions of central control, handles nominations (dispatch in relation with the flow management), and monitors the status of transmission system.

### 4.3.1 Manned and Unmanned Facilities

Basically, all the stations of the transmission system are fully automated for all essential functions and are remote monitored or controlled from the CCR.

#### *4.3.1.1 Manned Facilities*

**Central control room:** It is manned round the clock (24 hours) by teams operating in shifts wherein each team is of two operators. One operator is responsible for supervision of the system and the other one is responsible for flow management and relations with shippers and end users. However, both the operators are trained to fully manage both functions.

**Compressor stations:** They are fully automated and usually are monitored (variation in flow) or controlled (machines) from the CCR. However, local actions are needed from time to time for the unplanned start of machines and so on which requires manning of such facilities.

**Unmanned facilities:** All other stations (i.e., block valve stations, end users' facilities, and city gates) are unmanned from the view point of operator, only safety guards attend these stations permanently. These are fully automated (local remote terminal unit (RTU)/ programmable logic controller (PLC)) and remote monitored or controlled by or from the CCR.

Moreover, all stations are protected by anti-intrusion detection system which are equipped with closed-circuit television (CCTV) monitored from the central control room (CCR).

### 4.3.2 Central Control System

The remote control and monitoring is basically ensured from the main central control room located near the administrative office of the company. In the transmission line there is one main control room and there may be multiple secondary control rooms depending upon the need to monitor and control the operations across the transmission line. The secondary CCR continuously exchanges information with main CCR so that the main CCR remains updated with all the necessary information to take over the control in case of failure of the main CCR.

In the main CCR, two SCADA servers in dual hot standby modes are provided so that in case of failure of the active server a smooth change over (i.e., bumpless) is accomplished automatically to the standby server without affecting the SCADA functionality in not more than 3 minutes.

#### *4.3.2.1 Main Central Control Room*

The main central control room and related instrumentation are directly adjacent to the equipment assuring other functionality, which the transmission firm implements as part of central gas management system (CGSM). However, SCADA also must have the necessary provision to interface with various APPS modules such as (1) the on-line network simulator and leak detection and (2) the flow management and nomination handling.

Two SCADA servers are installed for redundancy to mitigate the risks. The system acquires and records the status of all main equipment which are part of the gas transmission system (GTS). The brief details are as follows:

1. Block-valve stations: The system at block valve stations monitors the position of main valves, gas temperature and pressure, alarms, anti-intrusion, and so on.
2. Entry point: At this point, status of main inlet and outlet valves, pressure and temperature, status of separator and filters, data from flow computers, alarms, and so on is monitored.
3. Delivery facilities: At this facility, status of the main inlet valve, pressure and temperature at inlet, status of separators and filters, operating data from pre-heating system, pressure and temperature up- and downstream of the HIPPS, status of metering lines, data from flow computers, alarms, and so on is monitored and controlled.
4. The operational main data from the compressor stations and alarms is monitored.

The operator on duty can act remotely on:

- The main inlet valve at the entry point and each end-user delivery facility
- The selection of metering lines at the entry point and at each end-user delivery facility
- The main valve at any block valve station
- The set-points at each HIPPS at the end user's delivery facility
- The set-points of the machines at the compressor stations and the stopping of a single machine; however, the starting of a machine must be done locally

The following is applicable for remote actuation of valves:

1. Remote actuation can be inhibited locally (with an alarm in the CCR) for maintenance purpose and so on.
2. Closure remotely is always permitted.

Opening remotely is interlocked with a differential pressure measurement to prohibit actuation when the difference exceeds acceptable values for the equipment. Thus, this prohibition prevents damaging the valve seat, destroying filter cartridges, going in over speed with ultrasonic meters, and so on.

#### 4.3.2.2 Metering Data Transmitted to the Central Control Room

In addition to all actual data (corrected values, energy quantities, and so on), all primary data is transmitted continuously or by batch to the CCR for verification of computation for reconciliation purpose, if required. Results from gas chromatography also are transmitted continuously. If required, the operator on duty can remotely pilot the CCTV cameras to investigate the reason for some specific alarms (anti-intrusion, fire, and so on) before deciding on measures to take according to operating procedures.

#### 4.3.2.3 Secondary and Back-Up Central Control Room

Besides its function for local operations at the site, the CCR also acts as back-up (secondary CCR). The specific equipment for a secondary CCR consists of (1) a data server equivalent to the two servers for the main CCR (the server will be in continuous relation with the data servers in the main CCR), (2) one work station similar to the work station in main CCR, and (3) a high speed data communication interface with fiber optic cables for telecommunication linked to main CCR.

#### *4.3.2.4 Other Functionality*

Besides the conventional equipment for the CCR as mentioned in the previous section, the central control system (CCS) also must offer additional functionality through specific equipment directly linked with the SCADA and adjacent to or integrated in the main CCR. The functionalities offered are (1) an on-line network simulator and (2) leak detection.

##### *4.3.2.4.1 Online Network Simulator*

A network simulator is installed for the following:

1. To assure real-time follow-up of system behaviors under actual pressure, temperature, and flow conditions
2. To analyze actual trends and alert operator in case of predictable difficulties
3. To assist the flow management system for handling nomination requests

The network simulator basically monitors the real-time actual status of the GTS. It collects data (pressure. temperature, flow, and so on) on an actual real-time basis from the network. The data is then compared with its model and forecast nomination. The analysis of data helps in advising operators about acceptance of nominations (in case of excess gas). It also gives alert signals in case of danger to the system.

##### *4.3.2.4.2 Leak Detection*

The considerable improvement in network simulators and models for network behavior combined with adequate software and sophisticated SCADA realistically helps in leak detection on the main trunk line. Based on the amount of data collected on a real-time basis and network simulators, the transporter provides for such a system as an alert for the operator in the CCR.

### 4.3.3 Flow Management

Flow management drives the actual operation of the GTS by interfacing each shipper (and its end users) with the technical constraints of the system about the actual status. It considers the contractual terms and conditions of the individual gas transmission agreement (GTA) with a shipper.

Flow management is a complex operation. It is computer aided by a flow management module that has a direct relation with the CCS and with the dispatch from shippers.

Basically, it takes care of the following functions:

1. Nomination management
2. Relations with shippers and between shippers and end users
3. Administrative assistance for invoicing

#### 4.3.3.1 Nomination Management

Nominations are regulated by procedures provided in each individual GTA between the shippers and the transporter. According to the GTA, each shipper transmits at periodic intervals and advanced notice from the end users forecasts the gas requirements. Based on forecast, the shipper requests specific gas quantities for transport during the next period.

At the time of receiving the demand from a shipper, the transporter acknowledges the receipt, compares it with the contract, and aggregates it with the request from other shippers, if any. The transmission firm then verifies the total quantity for transport with the actual network status. If it is acceptable, the transmission firm will accept it and provide confirmation to shipper and/or reconcile it with the contract and review it with the shipper. After acceptance, the transporter instructs the operator on duty for further operation.

Whenever a shipper requests excess gas, the transporter must check carefully the actual status of its network together with other acceptable nominations and then decide whether to accept delivery of excess gas.

Moreover, in the case of operating problems on its network, each shipper must be duly informed and requested to reduce off-takes, when needed, to save the system.

When in the pipeline network more than one shipper is connected, the system must check that supplies at entry points are in adequate balance with off-takes and, in case of unacceptable imbalances, the transporter must initiate the allocation procedure.

Finally, due to the compressibility of gas, line-pack management is of prime importance to assure adequacy and anticipation between receipt at entry points and off-takes at delivery points.

To achieve all the previous tasks, the flow management module receives real-time information about the status of the transmission system (pressure, temperature, and so on) combined with instantaneous quantities received at each entry point and delivered at each delivery point plus its own use of gas for generating power to run the compressor.

**ILLUSTRATION**

AGL (a natural gas distribution company) has about 61,000 domestic customers and about 500 industrial and commercial customers. AGL sources natural gas from different suppliers and uses the transmission line network of two companies. Rainbow Transport Limited (RTL) and Gujarat Transport Limited (GTL) are two transmission line firms which offer their services to AGL for transporting natural gas from suppliers to the AGL distribution network. AGL has eight off-take points from which gas from the pipeline of GTL (a natural gas transmission company) is transferred to the pipelines of AGL.

The GTA requires AGL to notify GTL of the maximum delivery rate daily. In the GTA there is a clause that stipulates the notice period for variation in daily gas flow nomination by AGL. For example, notification for (1) a 20 percent variation in maximum daily delivery rate (MDR)/ maximum off-take rate (MOR), a gas transporter (i.e., GTL) requires a 4-hour notice, (2) a 10–20 percent variation in MDR/MOR requires a 2-hour notice, (3) a less than 10 percent variation in MDR/MOR requires a 1-hour notice, and (4) any decrease in the rate of delivery or off-take requires a minimum 2-hour notice prior to commencing the decrease in the rate of delivery or off-take.

These time constraints were a challenge for nomination in the initial stages of operations for AGL because monitoring and control processes at the off-take points were manual which made the nomination process more difficult. Furthermore, when AGL received the gas through RTL's transmission line, which transports the gas of other suppliers, up to 10 percent variation in calorific value is noticed compared to supply from the other gas suppliers through GTL's transmission line. Due to the change in calorific value, the consumption pattern of customers changed which resulted in a new challenge for nomination. However, for natural gas transmission companies such as GTL and RTL, accurate nomination is very helpful for management of their operations and for balancing the transmission network.

To ensure that the nomination from AGL is accurate, GTL and RTL through GTA imposed a penalty on AGL through the penalty clauses. In the initial period, AGL had to pay a penalty of about 2.7 percent to GTL and RTL for inaccurate nomination, which reduced to zero within 1 year. The reduction was possible because AGL improved its processes to forecast the demand.

#### *4.3.3.2 Relations with Shippers and Between Shippers and End Users*

The CCS is the entity which receives all actual data from each entry point and each delivery point. Provision is made to allow shippers to collect such data from the CCS. The CCS is connected to the data transmission system developed by the transporter for its own use.

The CCS (and back-up room) has direct communication with shippers who are dispatching gas and the local control room of each end user to allow quick data exchange in case of emergency situations. Voice transmission is also provided.

#### *4.3.3.3 Administrative Assistance*

The CCS is connected to the administrative department of the gas transmission company to assist in preparing the invoices.

### 4.3.4 Supervisory Control and Data Acquisition and Telecommunication

Following is a brief description of the system:

1. The CCS continuously and in real-time monitors or remote controls each station (entry point, compressor station, block valve, and delivery facility) in the pipeline transmission system.
2. Telecommunication and data exchange are assured through a private fiber optic cable installed along the pipeline during construction.

The connections between the CCS and the fiber optic cable along the pipeline may be planned using specific data lines which are leased from local telecommunication operators.

Back-up links to essential and critical stations may be assured through modems and local telecommunication operators for which the modalities are planned at the project implementation stage.

As such, very limited fiber optic capacity is required for the transmission line project. It is suggested to install additional spare fiber optic lines in the backbone along the main pipeline so that spare capacity can be offered to third parties.

### 4.3.5 Geographical Information System

All the components are traced and surveyed geographically. A comprehensive geographical information system (GIS) may be developed such that it provides rapid information and location identification to operations and maintenance (O & M) staff. The GIS must provide the location of the pipeline to third parties intending to interfere with the system for works on and/or construction of other infrastructure.

The system may be conceived to enable direct information exchange with similar GIS which exists or may be developed at the national, regional, and local level in the country.

## Bibliography

https://primis.phmsa.dot.gov (accessed on June 20, 2018).
https://www.fractracker.org/2016/06/introduction-oil-gas-pipelines/ (accessed on May 29, 2018).
https://www.aga.org/natural-gas/delivery/how-does-the-natural-gas-delivery-system-work- (accessed on April 15, 2018).
http://NaturalGas.org/naturalgas/transport/ (accessed on May 15, 2018).
http://naturalgassolution.org/policy-issues/infrastructure/
Sudhir, Y. (2013). "Natural gas demand forecasting and nomination process of a gas utility company," *ICBM 2013 Bangkok Conference Proceedings.*
www.Amtechvalves.com.
www.balstonfilters.com.
www.controlplus.in.
www.pgjonline.com.

# 5

# *Natural Gas Distribution Business: Project Management Aspects*

The natural gas distribution business is the last mile connectivity to users of natural gas as a source of energy. The firms engaged in this business are also called natural gas utilities. Natural gas utility firms supply natural gas to various customer segments (i.e., industrial, commercial, transport, and households). This business requires firms to lay, build, operate, and expand the natural gas distribution network according to the policies of the local governments.

Constructing natural gas transportation networks in congested and populated city areas is a challenging task that involves not only capital, labor, and equipment, but also intense coordination with other utilities and agencies. These projects must be completed on time without an escalation of costs. The best practices of project management thus must be adopted by city gas distribution (CGD) firms. This chapter discusses the project management aspects of the natural gas distribution business.

## 5.1 Natural Gas Distribution System Description

The natural gas distribution network means an interconnected network of gas pipelines and the associated equipment used for transporting natural gas from a bulk supply high-pressure transmission main to the medium pressure distribution grid and subsequently to the service pipes supplying natural gas to domestic, industrial, or commercial premises and compressed natural gas (CNG) stations situated in a specified geographical area. The schematic diagram of a natural gas distribution network is shown in Figure 5.1.

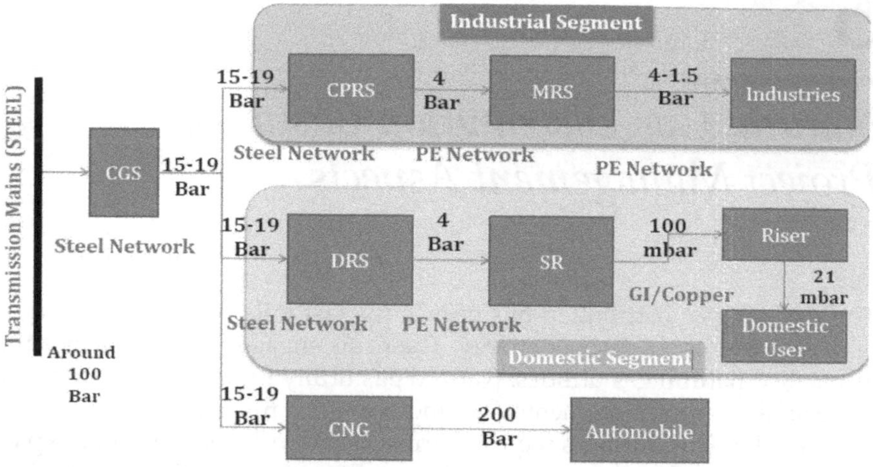


**FIGURE 5.1**
Schematic diagram of a natural gas distribution network-materials and pressures. (From Akbari, D., A Study of CGD Business, Vocational Training Report, 2nd Year Petroleum Engineering, PDPU.)

A natural gas distribution network system can be classified into four categories:

1. Primary network: A medium pressure distribution system comprised of pipelines, gas mains, or distribution mains usually constructed using steel pipes and connecting one or more transmission pipelines to respective city gate stations (CGSs) or one or more CGSs to one or more district pressure regulating stations (DPRSs). The maximum velocity in the pipeline network may be limited to 100 feet per second (30 meters per second) immediately after a pressure regulating instrument. As far as practical, a primary network must be fed through more than one CGSs or source of supply. The operating pressure must be as defined by the regulatory authority. It normally operates at pressures above 100 pounds per square inch gauge (7 bar) and below 711 pounds per square inch gauge (49 bar) and pipelines forming part of this network are called gas mains, distribution mains, or ring mains. The network is designed to ensure an uninterrupted supply of gas from one or more CGSs to supply gas to the secondary gas distribution network or service lines to bulk customers through the service lines.
2. Secondary network: A low pressure distribution system comprising of gas mains or distribution mains usually constructed using

thermoplastic piping (i.e., medium density poly ethylene (MDPE)) and connects a DPRS to various service regulators at commercial, industrial, and domestic consumers. The network is designed for maximum flow velocity of 100 feet per second (30 meters per second). It operates at a pressure below 100 pounds per square inch gauge (7 bar) and above 1.5 pounds per square inch gauge (100 mbar) and pipelines forming part of this network are called low-pressure distribution mains, which must be designed to ensure uninterrupted supply to tertiary networks or to industrial consumers through service lines.

3. Tertiary network: A service pressure distribution system comprised of service lines, service regulators, and customer or consumer meter set assemblies constructed using a combination of thermoplastic MDPE piping and galvanized iron (GI) and copper (Cu) tubing components. The network operates at pressure less than 1.5 pounds per square inch gauge (100 mbar) and pipelines forming part of this network service are pressure distribution mains, which must be designed to ensure uninterrupted gas supply to service lines.
4. Sub-transmission pipeline (also referred as Spur Lines) are high-pressure pipelines connecting the main transmission pipeline to the CGS. It is owned by the natural gas distribution firm.

A typical distribution network comprises of one or more or all the following:

1. CGS
2. Pipeline network: steel pipeline, polyethylene pipelines, GI/Cu pipes
3. Regulating stations: district regulating stations (DRS), service regulators (SRs), individual pressure regulating station (IPRS), domestic, commercial, and industrial regulators
4. Metering stations and metering and regulating stations (MRS)
5. CNG Stations

### 5.1.1 City Gate Station

The CGS is established at a tap-off point of the high-pressure transmission pipeline. The natural gas is drawn at the CGS from a transmission line and then it is fed into the distribution network by the distribution or utility firm. At CGS, the custody transfer of natural gas from the transmission company to the distribution company takes place. The gas delivered at this point is at higher pressure (i.e., greater than 50 bar). When gas enters the CGS, unit pressure is reduced to 24–30 bar. The main components of a CGS and their function are as follows (Figures 5.2 through 5.6):

1. **Filtration skid:** Dust particles and liquid mixed with the gas stream are separated by high efficiency filters. Up to the filtration skid the same pressure is maintained from the inlet to the filtration skid.

   After the filtration two streams are divided from the main line using a header. The line which is used is called the active line whereas another one is called the passive line.

**FIGURE 5.2**
City gate station skid. (From www.controlplus.in.)

**FIGURE 5.3**
Filtration skid. (From www. controlplus.com.)

2. **Pressure reduction unit:** A pressure reduction valve is installed for the reduction of the gas stream pressure from 60–65 bars to 25–30 bars. A creep relief valve and a slam shut off valve is installed in this skid for safety. A pressure reduction unit has a twin stream arrangement with one stream in working mode and the other kept in hot stand-by mode. The PRS unit also must have active and monitored pressure-reducing valves in tandem in both streams.
3. **Metering skid:** The metering skid is installed for gas flow measurement. Ultrasonic meter, turbine meter, or orifice meters can be used in the metering skid for gas custody transfer.

**FIGURE 5.4**
Pressure reduction skid. (From www.controlplus.in.)

**FIGURE 5.5**
Meter and custody transfer skid. (From www.controlplus.in.)

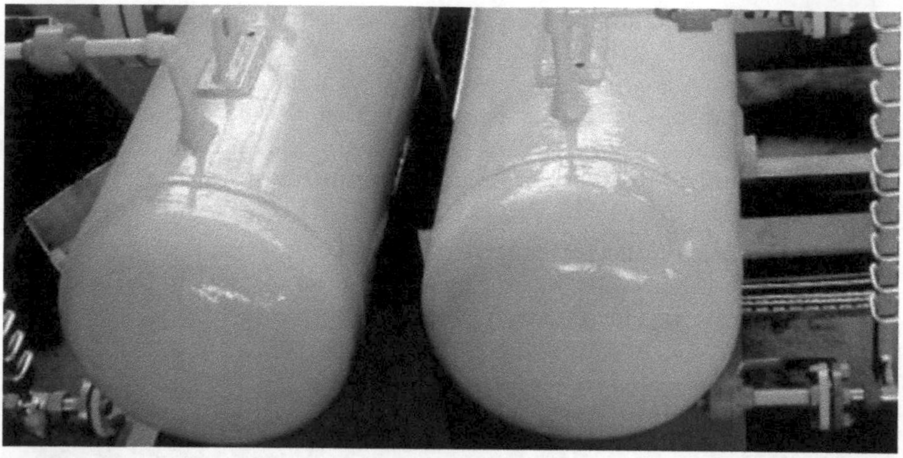

**FIGURE 5.6**
Odorizing system. (From www. controlplus.in.)

The various parameters such as temperature in the various sections of the line and pressure at the inlet and outlet flow rate are monitored remotely by the SCADA systems in the control room.

4. **Odorization unit:** An odorization unit is installed at the CGS to add an odorant (ethyl mercaptan or equivalent) in the gas stream. The odorant dosing rate is between 10 and 12 milligrams per meter cube according to the requirement. This unit consists mainly of two odorant 200 liter storage tanks, an expansion tank, a pneumatic panel, a level indicator, and a filter. Ethyl mercaptan boils at 35°C. To stabilize its liquid state, a high-pressure gas blanket is provided above it in a storage tank. A positive displacement pump is installed to dose a fixed volume of odorant per stroke. During pumping, dosing rate is calculated in the liquid state. The odorant unit works in a closed-loop system to avoid odorant spillage. During maintenance, the ethyl mercaptan is pumped to the storage tank by a process called purging. The odorization unit is directly connected to the main line with stainless steel (SS) tubing. The on-line odorization equipment is designed to minimize fugitive emissions during loading, operation, and maintenance.

### 5.1.2 Pipeline Network

As discussed in a previous section, the primary network is constructed of steel pipes, the secondary network is constructed of MDPE pipes, and tertiary network is developed from GI or Cu material. Spur lines are constructed of steel material.

### 5.1.3 Regulating Stations

1. **District Regulating Stations (DRS) or District Pressure Regulating Stations (DPRS):** DRS are the device used to reduce the pressure from 25 to 4 bars. It is the interface between the steel grid network and the medium pressure MDPE network. It is located at various demand centers for domestic and commercial users and typically consists of (1) gas filters, (2) heaters, if required, (3) pressure-reduction (active and monitor combination) skid with a minimum 50 percent redundancy including a slam shut valve for over pressure protection, (4) a creep relief valve, (5) pressure gauges, (6) non-return valve (NRV), (7) pilot regulators, and (8) inlet and outlet isolation valves.

   The flow capacity of DRS is in range 5000–10,000 standard cubic meters per hour (SCMH). The inlet line of DRS is steel pipeline and the outlet is MDPE pipeline (Figures 5.7 and 5.8).

**FIGURE 5.7**
District regulating skid. (From www.controlplus.in.)

FIGURE 5.8
Domestic and industrial skids. (From www.controlplus.in.)

2. **Individual Pressure Regulating Station (IPRS):** IPRSs are located on the premises of an individual customer. They have facilities like DPRS; however, monitor regulators may or may not be provided at the IPRS. Metering facilities may or may not be part of this station.
3. **Service Regulator (SR):** It is located on the customer's premises to maintain supply pressure and safe conditions in the event of rupture to regulate downstream section.

### 5.1.4 Metering and Regulating Stations

The function of an MRS is to monitor the flow rate, measure gas volume, and reduce the gas pressure to meet the requirements of gas pressure in the downstream. The gas volume measured at this location is used to raise the invoice or bill. These are installed at individual industrial customer's premises where the distribution company hands over the custody of gas to the end user.

### 5.1.5 Compressed Natural Gas Stations

CNG stations are located along with petro-retail stations or it may have an exclusive CNG station. CNG stations may have an on-line mother station or may have cascade with daughter booster stations (DBSs). It feeds the natural gas to vehicles through CNG dispensers. To ensure operation of the CNG station, an uninterrupted power supply is required (Figures 5.9 and 5.10).

**FIGURE 5.9**
Mother compressed natural gas station. (From www.gruppotecnogas.it.)

**FIGURE 5.10**
Mother compressed natural gas filling station. (From www.gruppotecnogas.it.)

## 5.2 Project Management

### 5.2.1 Project Planning: Natural Gas Distribution Network

The federal government through the regulatory authority or body develops and monitors the activities related to the natural gas distribution network. The regulatory authority of the respective country assesses the requirement of natural gas as a source of energy in various geographical areas (GAs) in line with the federal government policies. Based on the assessment and requirement of natural gas and business potential of natural gas as an energy source, the concerned authority invites expression of interest (bids) from business entities to build, operate, and maintain the distribution network for specific GAs.

The eligibility criteria to bid for the license for a natural gas distribution business may include experience of the firm in building, operating, and maintaining the network, adequate number of technically qualified personnel, net worth of the bidder, and so on. The bids may be evaluated using the present value based on the bid parameters like network tariff (yearly), compression charges (yearly) for CNG, and the number of CNG dispensing stations through economic life of the project (i.e., 25 years). Furthermore, the present value of investment in inch-kilometer of steel pipeline laid and numbers of domestic connections covered or offered during the period of exclusivity. Each of the parameters may be given appropriate weightage depending upon the requirement of the respective country. The entity that gets the highest composite score on the parameters is awarded the GA. After the GA is awarded the firm starts building the natural gas distribution network.

### 5.2.2 Project Analysis: Natural Gas Distribution Business

To prepare the bid documents, the interested firms must carry out a techno-economic feasibility study for the defined GA for which the firm wishes to bid to get the license. A techno-economic feasibility study includes (1) market analysis, (2) technical analysis, and (3) financial analysis.

#### *5.2.2.1 Market Analysis*

The market analysis of a defined GA is undertaken to assess the demand of natural gas and supply sources in the defined GA. Demand and supply is assessed for the economic life of the project which is considered 25 years for a natural gas distribution business. The demand estimates and supply sources or availability of natural gas is the base for the techno-economic feasibility of the natural gas utility business. Demand and supply are the critical parameter for the technical design and the financial feasibility analysis.

##### *5.2.2.1.1 Demand Analysis*

To assess the demand potential, the following parameters are considered: (a) switchover by existing users of energy from an alternate source of energy to natural gas based on economics, guidelines by statutory bodies, and so on, (b) economic growth in the GA (i.e., gross domestic product (GDP) and industrial growth), and (c) any other relevant factor (i.e., policy of government, sources of supply, or proposed connection of the distribution network to existing transmission networks).

Demand potential for the economic life of the project is worked for three scenarios (i.e., optimistic, realistic, and pessimistic). To assess the demand, the key stakeholders are consulted (which may be the anchor load industry, industry association, and large commercial complexes) for obtaining their thoughts on acceptance of natural gas as fuel in the region.

To estimate the demand, geographic, demographic, and economic data of the GA is collected. The information related to census data (household population) in the GA and its estimated growth over the economic life of the project, current use of energy type, and potential to switch over to natural gas is collected. The demand estimates also must be carried out for various segments (i.e., commercial, industrial, domestic, and transport). It is also suggested to consider the development plans of the local government and other development authorities, which identifies the major industrial area, industry parks, special economic zones (SEZs), development of industrial corridors, commercial zones, office or administrative blocks, bazaars, aerodromes, railway and bus stations, malls and plazas, housing estates, and any major energy-consuming areas in the GA. Statistical tools are used to forecast the demand. The various methods for forecast used are either qualitative or quantitative in nature. The quantitative methods used to forecast the demand of natural are multiple regression method, time series analysis, impact of seasonal factors in statistical analysis, and so on. Some of the qualitative methods used are experts opinion, end-use methods, and so on.

While forecasting the demand, various government notifications related to switching to natural gas energy for all four segments, plans to set-up mass rapid transport systems, development of state highways, national highways, and water ways, and so on are considered. All the assumptions related to demand forecast must be mentioned upfront in the analysis.

**ILLUSTRATION**

1. One of the natural gas utility companies started its operations and supplied natural gas as a fuel to one of the industrial units which was manufacturing steel alloys. The utility laid the steel pipeline and other pipeline network with a certain diameter to connect the steel manufacturing unit to the supply of natural gas. Within 2–3 years, the demand for natural gas by this industrial unit increased multifold. The increase resulted into the pipeline network capacity constraint (in terms of pipeline diameter) to meet the increased demand of the buyer and customer. It required that the utility company either lay a parallel infrastructure of pipeline or replace the existing pipeline network with pipes of a larger diameter. Thus, if the demand forecast is not reasonably accurate, then it will increase the capital expenses (CAPEX) of the utility company.

(*Continued*)

**ILLUSTRATION (Continued)**

2. In another instance the natural gas distribution company set up an on-line CNG station to serve the transport sector on the premises of the State Transport Corporation. It was envisaged that most of the state transport buses would be converted to natural gas-based engines. The demand increased to a level such that the utility had to increase the capacity of the compressor by installing two more compressors within 10 years of operations. Currently, the demand is very low, and the utility is unable to use its capacity because the number of state transport buses with natural gas-based engine reduced drastically. The cost of operating and maintaining the CNG facilities is higher compared to sales. Therefore, the utility firm is now planning to shift this infrastructure to another suitable location. Since infrastructure projects have a long gestation period, it is very critical to forecast the demand through the economic life of the project.

##### *5.2.2.1.2 Supply Market Analysis*

Supply market analysis refers to gas sourcing and pipeline connectivity analysis. All feasible gas sourcing options (i.e., domestically produced gas, regasified liquefied natural gas (RLNG), and so on) may be analyzed because the project has a long gestation period (i.e., 25 years). The factors to consider are government policy, affordable prices, the connectivity of the source of gas with the transmission line, liquid natural gas tank trucks, or any other mode. A natural gas source is evaluated for its potential over economic life (i.e., 25 years), quantity, price, location, and supplier to meet the requirements of natural gas distribution. It is required to consider also the existing transmission line connectivity and future proposed transmission line passing by the GA.

#### *5.2.2.2 Technical Analysis*

Natural gas distribution network definitions, design, materials and equipment, welding, piping system components and fabrication, installation, testing, operation and maintenance, corrosion control, and other miscellaneous matters must be in accordance with requirements of the American Society of Mechanical Engineers (ASME) B31.8 and the regulations laid down by the regulatory authority of the federal government. The selection of design for the natural gas distribution network is based on the gas properties, required flow rates, operating pressures, and the environment.

##### *5.2.2.2.1 Design Philosophy*

The philosophy for the system design is worked out considering the availability of enough gas in the CGSs which meets the demand for the entire area. While designing, the factors to consider are (1) immediate and long-term business interests of the firm, (2) the overall economics of the available options considering the large expanse of the GA, and (3) minimizing dead length of the pipeline because the total area of GA may be very large.

The design of natural gas distribution networks and components must ensure (1) supply of gas at a constant volume into a system, which may fluctuate in pressure between pre-determined upper and lower limits in the distribution network, (2) supply of gas at a constant pressure at consumer end, and (3) the design must recognize the need for a safe guard against malfunction of any equipment and provide sufficient redundancy to ensure that the supply is secured against such malfunctions.

Facilities forming part of natural gas distribution networks must be designed considering (1) range of flow rates and pressures required in various sections of the network, (2) quality of gas, including cleanliness in respect of both solid and liquid particles, (3) metering requirements, (4) noise control, and (5) corrosion protection. Necessary calculations must be carried out to verify the structural integrity and the stability of the pipeline for the combined effect of pressure, temperature, bending, soil/pipe interaction, external loads, and other environmental parameters, as applicable, during all phases of work from installation to operation.

The aspects for consideration in deciding the layout of facilities at CGS, DPRS, IPRS, and so on are (1) type and size of equipment to be used (2) piping and instrument diagrams (P&IDs), (3) utility requirement, (4) venting, if required, and (5) operation and maintenance of the facility.

The augmentation of infrastructure for CNG and piped natural gas (PNG) distribution can be planned in a phased manner as the demand builds up slowly in subsequent years. It would ensure capital investment in phased manner resulting into overall economy of the project. As the area covered may be quite large with many locations to reach, the philosophy, which may be adopted for the CNG system design, is to have centralized CNG compression facilities at few locations. Other locations, comparatively farther away from the steel network, may be fed with DBSs. The compression capacity requirements for these DBS locations must be factored in for the CNG mother stations planned along the steel network. It is aimed at optimizing the steel pipeline network which may yet have a good geographical spread and area coverage of CNG availability.

The network must be conceived to be able to cater to the peak load realizable demand for the 25th year. It should avoid the problem of having to lay another line on the same route after the exclusivity period. The CNG dispensing facilities installation may be proposed in a staggered manner which should be in line with the projected realizable demand for future relevant years. Similarly, for the domestic, commercial, and industrial sectors, the installation of MDPE pipelines and so on may also be planned in stages.

#### 5.2.2.2.2 *Source of Gas*

Natural gas for supply in the GAs is to be taken from trunk lines or spur lines. Since the GA to be served may be very large (multiple towns and cities, various industrial areas, and so on) with considerable distance between them, the firms need to plan multiple tap-off points from the Trunk Lines and Spur Lines to source the gas. It may not be economically viable to conceive a network in the complete area (the entire GA) with a single or couple of sources and reaching all areas from those few sources. It is advisable to have multiple sources of gas from points on the source pipelines suitably located for major demand centers.

Thus, the distribution network is conceived in such a way that the total network in the GA essentially comprises a few smaller gas distribution networks which may be, for all practical purposes, independent of each other. Thus, independent networks would avoid the requirement to interconnect them and would result in eliminating the dead length of the steel pipeline.

Thus, for optimization of the pipeline network and downstream distribution, the GA may be divided into distinct areas and for each area tap-off points may be planned to source the gas. The sourcing point may have CGS, SV stations, IP stations, and so on.

CGS may be planned at multiple locations of gas sourcing and equipped with metering skids of the required capacity. Natural gas at an outlet pressure between 24–49 kg/cm$^2$(g) at the CGSs must be available for the PNG and CNG facilities in the GAs.

As mentioned previously, the odorizing units with ethyl mercaptan as odorant are installed at various CGSs in line with the regulatory requirement.

#### 5.2.2.2.3 *City Gate Station*

The CGS may be installed at the periphery of populated area. As mentioned previously, it is advisable to have more than one CGS for supply security. Properly laid out roads around various facilities must be provided within the installation area for smooth vehicular access. The facility must be safeguarded with the proper boundary wall and fencing with one or more gates in line with the regulatory authority and federal government guidelines.

Various facilities with buried pipeline, coating on the pipeline, a venting facility, gas detectors, fired gas heaters, and so on within the CGS area must be in line with the regulatory requirements. It must be ensured that the operations and maintenance of all the equipment and piping is easy and convenient.

#### 5.2.2.2.4 *Distribution Pressure Regulating Station (DPRS) and Individual Pressure Regulating Station (IPRS)*

DPRS facilities can be located above ground or below ground according to the regulations. IPRS usually must be located above ground. DPRS and IPRS installed above ground must be provided with proper security fencing according to the requirement of the local authorities. The distance between the fencing and the wall of the nearest building or structure must be according to the norms of the regulatory authority to ensure safety.

*5.2.2.2.5 Design of the Pipeline and the Compressed Natural Gas Network*

*5.2.2.2.5.1 Field Engineering and Survey* Field engineering and survey involves planning the locations of CGS in GAs, the most optimal route, and other related infrastructure for natural gas distribution. The activity involves mapping and phasing of required infrastructure, equipments, and facilities based on demand in different segments over the economic life of the project. While doing this, due consideration is given for peak load requirement, redundancies, and contractual obligations for booking the pipeline capacity for sourcing gas and its transmission until reaching the CGS. The design is based on approved technology and practices laid-down by the concerned regulatory authorities. The technical and safety standards including specifications of the regulatory authorities must be complied with.

Approximate bill of material (BOM) is prepared based on a mapping and phasing exercise which includes (a) the steel pipeline network layout design according to concerned technical standards of the regulatory authority, (b) estimating the cost of each of the phases with details reflecting the cost of the facility or equipment-type with break-up of material and labor, laying and erection cost, and so on, (c) identify and evaluate the design alternatives with a cost-benefit analysis adopting the relevant and latest technology.

The main objective is to configure the distribution network in the most optimal manner and estimate the CAPEX and operating expense (OPEX) and resource deployment plan during the economic life of the project. Redundancies of equipment and supply (i.e., ring-mains), reliability of equipment and services, and health, safety, security, and environment (HSSE) issues are clearly highlighted with cost-effective solutions. Large areas are highlighted in the GA and all issues related to technical, commercial, and HSSE terms including crossing over bodies and climatic conditions are addressed appropriately.

*5.2.2.2.5.2 Peak Load Design* The distribution pipeline network is designed for the peak load flow rate. The rate of consumption of gas in various consumer sectors (i.e., market segments) is not uniform. As such, the maximum quantity of gas for a day in each sector or segment is consumed within a fixed time in a specified manner. During this period, the rate of flow of natural gas increases. This increased flow rate is termed as peak load. The system must be designed for a higher capacity based on the peak load requirements of all consumer segments. The peak loads are calculated to design the system which is based on the realizable demand up to the 25th year or the life of the network. The peak loads are worked out using the following assumptions for each segment:

1. CNG segment for automobile sector: The peak loads are worked out considering 60 percent of daily demand to be met in 8 hours (daytime) and the balance of 40 percent in remaining period.
2. PNG segment for industrial sector: The peak loads are worked out considering 18 hours working in industrial units.

3. PNG segment for commercial sector: The peak loads are worked out considering 12 hours working in commercial units.
4. PNG segment for domestic users: The peak loads are worked out considering consumption in 8 hours (daytime) per day.

*5.2.2.2.5.3 Steel Pipeline Design* The steel grid line networks are designed to cater to a pressure up to 49 kg/cm$^2$; however, the maximum working pressure is considered to be 24 kg/cm$^2$. As mentioned previously, multiple CGSs are planned for supplying gas to the various areas in GA. These independent CGSs feed gas to the different pipeline systems which are for all practical purposes independent of each other. All these pipeline systems, put together, traverse all over the city area from the various CGS. It is planned in phases depending upon the growth of market. The distribution network system is developed by optimizing the pipeline sizes using engineering design software under various demand scenarios. The pipeline may have the following design parameters:

1. Sizes used: 4-, 6-, and 8-inch diameter steel pipes with appropriate thickness based on design parameters
2. Design pressure: 49 kg/cm$^2$(g)
3. Specific gravity of natural gas: 0.63
4. Density of gas: 0.76 kg/cm$^3$
5. Design temperature–buried and above ground according to the regulations and local climate
6. Joint factor: 1
7. Temperature factor: 1
8. Population density factor-grid line: Class–IV
9. Design life: 30 Years
10. Pipeline specification–pipeline material: for all sizes of pipeline: API 5L Gr. X–42/API 5L Gr. B 6.4 mm WT
11. Corrosion allowance: 0.5 mm
12. Pipeline Efficiency: 0.9

*5.2.2.2.5.4 Compressed Natural Gas Stations and Design Parameters for Compressed Natural Gas Stations*

**CNG Stations:** The CNG stations are planned in line with growth in demand for CNG. The CNG facilities are added in a staggered manner in phases. The on-line station/DBSs must be chosen so that they provide a good geographical spread for the CNG network.

CNG cascades of 3,000 water liters capacity are planned at all the Green Field mother stations. Adequate numbers of cascade along with light commercial vehicles (LCVs) are deployed to ensure uninterrupted CNG supply to DBSs. Each CNG station is planned with an adequate number of bus, car, and auto dispensers to maintain a healthy dispensing capacity to the gas demand ratio.

Capacity of a compressor at CNG stations may be of 1,200 SCMH. CNG compressors are also installed in phases to meet the growth in demand. CNG stations are set up either as a dealer owned dealer operated (DODO) or company owned company operated (COCO) model. The compressors usually are operated at 14–19 Kg/cm$^2$ to 255 Kg/cm$^2$.

**Design Parameters for a CNG Station:** The CNG stations are designed using 25 years (economic life of the network) of demand forecast. The number of compressors will be added over the years to maintain a healthy compression capacity for the gas demand ratio. Some of the indicative parameters used for design are as follows:

a. Average filling capacity: bus: 80 kg; car: 8 kg; auto: 3.5 kg
b. Average filling time/vehicle: bus: 8 minutes; car: 4 minutes; auto: 2 minutes
c. Number of arms per dispenser: bus dispenser: 1; car dispenser: 2; auto dispenser: 2

*5.2.2.2.6 Design Parameters for a Piped Natural Gas Distribution Network*

The pressure norms that may be followed for the PNG distribution network system design are as follows: (1) distribution/service line (medium pressure system): 6–1.5 bar, MDPE installation, (2) domestic connection (low pressure system): 21 mbar, GI/ Cu installation, (3) supply pressure large commercial consumer: 2 bar, (4) supply pressure small commercial consumer: 300/500 mbar, (5) supply pressure for industrial consumers: 2 bar or according to requirement, and (6) supply pressure to commercial consumers may be need based.

The industrial units in different areas may be catered through the PRSs or MRSs meant for the area. Individual industrial units must be supplied through a steel pipeline or an MDPE network originating from the DRSs, depending on the load requirements.

The commercial consumers like hotels, restaurants, food joints, hostels, and hospitals are catered to through an MDPE network which may be through PRSs or DRSs.

The domestic connections would be provided with uninterrupted gas supply at a pressure of 21 mbar through PRSs or DRSs.

*5.2.2.2.6.1 Pressure or District Regulating Stations (PRS or DRS)* It usually is planned to have common PRSs or DRSs (as the case may be) for all domestic and commercial connections and for some of the industrial connections. The capacity of DRSs and PRSs planned may be of 600, 1,000, 2,000, and 5,000 SCMH.

*5.2.2.2.6.2 Medium Density Poly Ethylene Network* A MDPE network planned for a natural gas distribution pipeline may be of sizes such as (1) diameter: 180 mm, PE 100, SDR 11, (2) diameter: 125 mm, PE 100, SDR 11,

(3) diameter: 63 mm, PE 100, SDR 11, (4) diameter: 32 mm, PE 100, SDR 11 and (5) diameter: 20 mm, PE 100, SDR 11.

The MDPE network begins from an MRS or DRS and enters the premises of industrial, domestic, and commercial units.

*5.2.2.2.7 Technical Standards for Natural Gas Distribution Network (Codes, Specifications, and Guidelines)*

The federal government or its authorized regulatory body develops and implements the technical standards including safety standards to build and operate the natural gas distribution network. The technical standards cover design requirements, standards for material and equipments, welding, piping system components and fabrication, installation and testing, operating and maintenance procedures, corrosion control, and other miscellaneous matters.

The material standards for steel pipes, MDPE pipes, Cu/GI pipes, and the CNG system are following:

1. **Steel pipes**

   The codes, specifications, and guidelines used for steel pipelines follow:

   a. ANSI/ASME B31.8
   b. OISD 141
   c. API 5L
   d. New pipeline policy including latest regulations of concerned authority

2. **MDPE pipelines**

   The codes, specifications, and guidelines used for MDPE pipelines follow:

   a. ANSI/ASME B31.8
   b. IS 14885/ISO: 4431
   c. OISD 220
   d. Latest regulations of the concerned regulatory authority

3. **Cu/GI pipelines:** According to standards laid down by the concerned regulatory body

4. **CNG system**

   The codes, specifications, and guidelines used for the CNG system in India follow:

   a. OISD (Oil Industry Safety Directorate): 179, 132, 110, 137, and 220
   b. NZS–5425
   c. NFPA–37, 52, 70
   d. ANSI B31.3 & B31.8

5. IS-2148
6. API–11P, API–618
7. Indian electricity rules
8. Explosives Act/regulations of concerned country
9. ASTM, National Electrical Manufacturer's Association (NEMA), Natural Gas Vehicle (NGV), National Electrical Code (NEC)
10. Chief Controller of Explosives (CCOE) & latest regulations of concern regulatory authority

Some of the other standards which may be available as a part of regulations are listed below:

(1) Design standards for steel pipes, MDPE pipes, CNG system, Cu/GI pipes, CGS, DPRS, IPRS stations, and so on, (2) electrical equipment and installations (3) instrument and control system, (4) buildings intended for human occupancy and location classes for design and construction, (5) design factors and location classes for steel pipe construction, (6) pipelines or mains on bridges, (7) metering facilities, (8) pressure/flow control facilities, (9) protection of pipelines and mains from hazards, (10) cover, clearance, and casing requirements for buried steel pipelines and mains, (11) clearance between pipelines or mains and other underground structures, habitable dwelling and industrial building, (12) casing requirements under railroads, highways, roads, or streets, (13) bends, elbows, and meters in steel pipelines and mains, and (14) hot taps.

Regulations also prescribe testing and commissioning standards which include standards for testing equipment and instruments, test required to prove strength of pipelines and mains to operate at specified minimum yield strength of pipe, testing requirements for DPRS and IPRS piping, hot tapes, safety practices to be followed at the test area, and so on. Test procedure followed may be according to ASME B 31.8.

Fire-fighting equipment must be available during commissioning. Proper communication facilities and a proper emergency response plan must be in place with emergency contact numbers of relevant agencies available at the testing and commissioning site.

Other regulations which must be followed are related to (1) plastic pipe and tubing joints and connections, (2) control and limiting of gas pressure in high-pressure steel, ductile iron, cast iron, or plastic distribution systems and in low-pressure distribution systems, (3) valves installation, and (4) vaults accessibility and so on.

The standards for gas service lines also may be referred which include general provisions applicable to steel, copper, and plastic service lines. It describes standards for installation of service lines, type of valves suitable for the service line, steel service line design, installation of service lines into or under buildings, design and installation of ductile iron service lines and plastic service lines, service line connections to mains, service line connections to steel mains, service line connections to plastic mains and so on, and piping beyond the consumer meter set assembly to the gas appliance.

### 5.2.3 Financial Analysis

The financial analysis is carried out based on the demand estimation in the GA, sourcing of natural gas, field engineering survey, and cost estimation of the distribution network design and infrastructure design. The capital cost, operating cost, and the purchase price of the gas delivered at the tap-off and the sales price finally to consumers of the natural gas for the project helps in working out the projected profit & loss statement, cash flow statement, and balance sheet for the project life of 25 years. Furthermore, the analysis must also explore the means of financing (i.e., equity and debt), the internal rate of return (IRR) of the project, the net present value (NPV) of the project, financial ratios and margin analysis, working capital requirements, and so on for the project life of 25 years.

#### *5.2.3.1 Project Cost Estimates*

Capital cost is worked out based on creation of facilities to serve the market. It includes CGS and SVs, steel pipe grid line, CNG stations, and the CGD network (of MDPE pipes with varying diameter) for domestic, commercial, and industrial connections. The aggregate capital cost breakup (as a percent of total cost) for the natural gas distribution business is given in Table 5.1.

**TABLE 5.1**

Cost Estimates for a City Gas Distribution Project

| Description | Cost (%) |
|---|---|
| Survey Work | 2 |
| CGS & Steel Pipeline Network (Material and Laying) | 23.4 |
| Last Mile Connectivity for Industrial Users | 4 |
| Ball Valves | 0.1 |
| Odorizing Unit | 0.5 |
| CNG Stations (CNG Compressors, CNG Cascades, CNG Dispensers, CNG Composite works) | 23 |
| MDPE (Material and Laying) | 25 |
| Metering Skids | 7 |
| Smart Meters | 6 |
| ROU Charges to various authorities | 4 |
| Land Cost for CGS, DRS, CNG Stations etc | 5 |
| TOTAL | 100% |
| EPMC Fees[a] | 5 |
| Owner's Management Expenses[a] | 7.5 |
| Contingency[a] | 5 |

[a] Aggregate cost shows % of total project cost

The approximate cost for various units follows (as of year 2017/2018):

1. CNG stations (construction work): USD 400,000 per station
2. CNG compressors (gas driven) of capacity 1,200 SCMH: USD 250,000 per compressor
3. CNG cascade: USD 16,000 per cascade
4. CNG dispensers: USD 15,000 per dispenser
5. Steel pipeline (material + construction): 8-inch diameter: USD 35,000 per kilometer; 6-inch diameter: USD 27,000 per kilometer; 4-inch diameter: USD 15,000 per kilometer
6. MDPE pipeline (material + construction): 125 mm diameter: USD 8,000 per kilometer; 90 mm diameter: USD 5,000 per kilometer; 63 mm diameter: USD 4,000 per kilometer; 32 mm diameter: USD 2,500 per kilometer
7. Metering skid: USD 2,000 to USD 12,000 per skid depending upon skid capacity ranging from G25 to G400

#### *5.2.3.2 Operating Cost*

The annual operating cost for the project at 100 percent capacity is worked out. Annual operating cost may include variable cost and fixed cost.

1. Variable costs: It includes cost of power, odorant, and natural gas.
2. Fixed costs: It includes cost of manpower, overheads, repair, maintenance and insurance, and annual maintenance contract (AMC) for compressors.

The various parameters considered for computing the operating costs follow:

1. Cost of energy: Electric energy cost is considered for CNG stations and for operation of electric motor-driven CNG compressors. In the case of gas engine-driven CNG compressors, the gas cost is considered equal to the purchase price of gas for the corresponding year.
2. Water: Since the water requirement of the project would be negligible, no separate cost towards it may be considered in the operating cost estimate.
3. Manpower: Annual cost of regular and contracted employees based on a yearly requirement may be worked out.
4. Repair and maintenance: Provision for repair and maintenance including replacement of infrastructure over the economic life of the

project may be kept at the rate of 1 percent of the capital cost for steel grid line and CGD and 5 percent for CNG stations. AMC may also be taken in addition to 5 percent of the capital cost taken for repair and maintenance of the compressors.

5. Insurance: Insurance cost at the rate of 0.25 percent of the capital cost may be considered in the operating cost estimates.

Further assumptions and parameters considered for financial analysis follow:

1. The project cost may be financed in a debt: equity ratio of 70:30 or according to the guideline of respective federal government. The interest rate and the repayment schedule and moratorium period for long-term loans may be considered, as applicable.
2. Provision for connection charges may be made in the CAPEX under two heads, namely MDPE network and domestic meters and regulators for yearly expected numbers of domestic connections. In case of commercial and industrial connections, it may be assumed that full expenses towards connection will be realized from the consumers. An interest-free deposit of a certain amount per connection for domestic consumers may be considered as a refundable advance. This amount usually is refunded at the end of project life.
3. Capital cost may be worked out based on the facilities envisaged for the steel grid line, CNG stations, and CGD networks including charges for domestic, commercial, and industrial connections including interest during construction. This capital cost is phased out based on the implementation schedule.
4. Yearly sales of gas volume (in SCM/MMTU) may be considered for analysis.
5. Gas purchase prices (yearly) at the tap-off point and by segment net-selling prices (yearly) for various segments (i.e., domestic, industrial, and commercial) and CNG is considered to work out the financial analysis of the project.
6. To arrive at the net-selling prices, prevailing taxes and duties are considered.
7. The working capital requirement for the project may be considered equivalent to 3 months of operating expenses.
8. Project life may be considered as 25 years from the date of authorization or award of the project.
9. Salvage value equivalent to 5 percent of the project cost may be considered at the end of the economic life of the project.

#### *5.2.3.3 Scenarios Building and Sensitivity Analysis*

The previously listed parameters can be worked out further for scenario building and sensitivity analysis.

1. A scenario building exercise is carried out to build various scenarios for risk analysis by examining the impact of various costs, prices, economic parameters, taxes and duties, supply risk, and competition risk.
2. A financial model is developed incorporating estimated demand of natural gas, CAPEX, OPEX, price of natural gas, and all other relevant parameters.
3. Sensitivity analysis using various critical parameters is undertaken and ROI, IRR, pay-back period, and so on is worked out.

Using the base case sensitivity analysis may be carried out to analyze the discussed parameters. The variation may be worked out as follows:

1. 10 percent increase or decrease in CAPEX
2. 10 percent increase or decrease in OPEX
3. 10 percent increase or decrease in gas volume sales
4. 10 percent increase or decrease in gas purchase prices
5. 10 percent increase or decrease in gas sales prices

The scenario building exercise and sensitivity analysis will help evaluate the risk in the project.

### 5.2.4 Project Implementation

The overall project comprises of the CNG and facilities including laying of the main grid and the MDPE network in the authorized GA. It is implemented over a period of 25 years/life of the project in a phased manner which accounts for yearly planning of the infrastructure development in line with the growth in demand for PNG and CNG in various segments. The targets for 3, 5, 10, 15, and 20 years are planned at the concept stage of the project so that it can be implemented smoothly. Some of the salient points for effective project implementation follow:

1. The development of the MDPE network is an activity which continues during the entire period/life of the project (i.e., 25 years). Furthermore, to achieve the target of providing the maximum possible number of domestic connections right from the first year, the simultaneous working for development of the MDPE network and main grid line (steel pipeline network) is envisaged.

2. Adequate infrastructure to cater to CNG requirements is planned considering the peak load demand. It requires installation of enough compressors, dispensers, cascades, and so on to ensure smooth CNG dispensing.
3. The steel grid line of 8, 6, and 4 inch diameters to connect CNG stations along with the required number of DRSs is planned for completion in the first year or next few years.
4. The implementation plan must be such that the network is developed within the period of exclusivity according to the regulatory authority guidelines.

The set of activities involved in development of city gate station and city grid line, CNG stations, and CGD follow:

1. **Construction of city gate station and city grid line**

   (a) Route survey of pipeline, (b) right of use (RoU) acquisition for the gridline, (c) design of facilities at city gate station proposed at tap-off point location, (d) basic and detailed engineering, (e) procurement of items for the grid line, (f) award of work contract for the grid line, (g) construction of the grid line, and (h) commissioning of the grid line.
2. **Construction of CNG stations**

   (a) survey the plots and existing oil marketing companies or retail outlets, (b) basic and detailed engineering, (c) procurement of compressors and other related items, (d) statutory approvals and land acquisition, (e) award of work contract for CNG stations, (f) construction of the boundary wall, building, canopy, and so on, (g) dispenser erection, (h) compressor erection, (i) cascade erection, (j) electrical works, and (k) commissioning.
3. **City gas distribution**

   The CGD work in different phases primarily comprises of installation of the required numbers of DRSs and development of the MDPE network. The construction of the CGD network includes the following activities:

   a. Survey and RoU acquisition and permissions
   b. Basic and detailed engineering
   c. Award of the work contract
   d. Procurement of items
   e. DRS erection
   f. Laying of MDPE pipeline
   g. Laying of GI/Cu pipe
   h. Installation of meters and regulators
   i. Testing and commissioning

#### *5.2.4.1 Critical Success Factors for Project Implementation*

1. All the required statutory approvals must be obtained according to the planned schedule.
2. RoU for laying a city grid pipeline route and land acquisition for the CNG station, DRS, and SV station, and so on is obtained according to the planned schedule.
3. Installation of CNG stations at retail outlets must be completed in time according to the time schedule.

The implementation schedule is planned after finalization of the engineering, procurement, and construction (EPC) consultant and the date of finalization of the EPC consultant may be considered as the "ZERO" date.

The firm must obtain the necessary technical services from an EPC consultant in the areas of basic design, detailed engineering, procurement of major equipment and materials, selection of suitable vendors and contractors, construction supervision, commissioning assistance, and project management.

The city grid line construction, erection of CNG stations, and laying the CGD system may be given to competent contractors to ensure completion of work according to the project schedule. All major and critical equipment and materials must be procured by the firm directly, as it ensures the quality of equipment and material as well as its availability in time.

## Bibliography

Akbari, D. A Study of CGD Business, Vocational Training Report, 2nd Year Petroleum Engineering, PDPU.

Al-Fattah, S. M. (2006). "Time series modeling for U.S. natural gas forecasting," *E-Journal of Petroleum Management and Economics*, 1(1), 1–17.

Carton, H. A. (2000). "Daily demand forecasting at Columbia gas," *Journal of Business Forecasting Methos & Systems*, 19(2), 10–15.

Chase, C. W. (1994). "Forecaster's view point," *The Journal of Business Forecasting Methods & Systems*, 13(1), 23.

Fildes, R., A. Randall, and P. Stubbs. (1997). "One day ahead demand forecasting in the utility industries: Two case studies," *Journal of the Operational Research Society*, 48, 15–24.

https://www.bharatpetroleum.com/pdf/EMPANELMENT.pdf (accessed on May 4, 2018).

www.controlplus.in.

www.controlplus.com.

www.gruppotecnogas.it.

http://www.pngrb.gov.in/pdf (accessed on July 6, 2018).

# 6

# *Natural Gas Distribution Business: Operations and Maintenance Aspects*

After the natural gas distribution network is set up and gas starts flowing in the pipeline network, the firm that owns the natural gas distribution network require developing the processes and creating an organization that can manage Operations and Maintenance (O&M) efficiently and effectively. The O&M processes must ensure continuous availability of gas to the consumers and maintain the health of the pipeline system. It must meet any emergencies arising in the system. The safety of the network system and the nearby areas is of utmost importance in the city area. Even a minor leakage or an accident or fire in gas pipeline may lead to a big disaster. Good O&M management helps to minimize the loss in production and property because of accidents.

There are more than 20 threats to the gas pipeline networks that are identified under the American Society of Mechanical Engineers code ASME B31.8S. These threats must be handled at the appropriate stage and time or it may cause a serious damage to the pipeline network. The network must be examined against each of the possible threats to identify the possible and potential hazards to the pipeline network. Accordingly, appropriate steps are taken to handle these threats right from the design and construction stage to the O&M stage for the entire life cycle of the pipeline network.

## 6.1 Operations & Maintenance Policy and Philosophy for Natural Gas Distribution Business

Natural gas distribution firms must develop and implement O&M policy to manage their business effectively and efficiently. A brief discussion about policy implementation follows:

1. **Compliance with O&M requirements**

   The application of the best O&M practices for the natural gas distribution system help firms to ensure safe and uninterrupted gas supply, integrity of network, and minimizes risk to personnel and property. The contractors or outsourced entity need to comply with the requirements of O&M of facilities and equipment, as practiced by the firms and its other engineering standards.

2. **Operating Standards and Instructions**

   Compliance with O&M standards and instructions must be enforced consistently for all the operations personnel because flaws and risks cannot be eliminated through design are controlled by operating standards and instructions.

3. **Deployment of Contractor**

   The natural gas distribution firm, if it wishes, may engage a competent contractor for O&M management of the natural gas distribution system to ensure safe operations, integrity of network, protection of those involved, and protection of environment and public at large. The contractor selection must be based solely on experience in a similar job and must have the required resources and capabilities.

4. **Inspection**

   Concerned personnel must conduct inspections periodically to detect and correct unsafe practices and conditions.

5. **Education and Training**

   All employees must be educated and trained to help them develop those skills needed to perform, supervise, and manage assigned tasks without mishap. The training must be exhaustive including various job skills and Health, Safety, Security and Environment management, especially, on the job and off the job safety, emergency handling, importance of personal protective equipment (PPE), and so on.

6. **Resources, Basic Facilities, Infrastructure, and Motivation**

   The natural gas distribution firm, in case of outsource activity, may provide the required resources, basic facilities, and infrastructure according to the contract term, to enable the contractor and its employees to carry out O&M activities in a safe environment. Good communications, a viable suggestion system, and the recognition of good O&M performance motivates contractors and employees to participate in effective O&M management programs.

7. **Job Placement**

   The utility firm must ensure that the contractor and employees will be assigned only tasks that are consistent with their capacities and job skills; this enables employees and the contractor to work safely and effectively.

8. **Response to Accidental Occurrences**

   The site specific effective emergency must be handled according to the emergency response plan (ERP) and according to the firm's guidelines. The guidelines may include measures to contain or control an emergency or disaster when an accident occurs to minimize the loss of resources. The guidelines may also include a reporting and investigation system to determine the cause of the accident, and the adoption of corrective actions to avoid a recurrence.

9. **Contractors' Employee Safety**

   The firm's official may monitor O&M management to ensure that the activities are performed by the contractor in conformity with O&M policies, statements, and practices and do not violate the set safe practices and procedures.

10. **Accountability**

    In case of contracted activity, the contractor and all the employees must be held accountable for personal and functional O&M performance. An important factor in the overall job performance evaluation of a contractor and employees may be how well the contractor and employee meets skill and safety responsibilities.

11. **Compliance Reviews**

    On a selective basis, teams that include people with related expertise to determine compliance with the policy may conduct compliance reviews. To ensure the credibility and effectiveness of the review, it must be conducted by an independent team.

## 6.2 Geographical Information System-Based Asset Management System and Mapping

To have safe and efficient operation of the pipeline system, availability of up-to-date systems maps and drawings is very important. The entity operating the natural gas distribution network must put in place a geographical information system (GIS)-based system that captures the entire underground gas network and customer database. All the pipelines laid by the natural gas distribution firm must be identified in GIS through geo-referenced coordinates. The failures in the pipeline network must be mapped in GIS for investigations. All the maps and drawings must be incorporated and maintained in the GIS mapping system. It must be accessible at all times specifically in the event of an emergency. In case of pipeline network modification or repairs, the details must be forwarded to the GIS group after completion of the construction or repair so that system maps and drawings may be updated. Hard copies of maps and drawings may be printed from the electronic files when required. Any changes to maps and drawings must be approved and communicated through the procedures of firm.

The information as a part of mapping process may contain (1) a list of pipeline segments, (2) pipe attributes (material of construction, length and type of pipe, and so on), (3) operating pressures, (4) maximum allowable operating pressure (MAOP), and (5) class locations.

The GIS-based system may have characteristic such that (1) the entire network may be viewed on one platform to manage the huge database, (2) all network extensions and expansions may be mapped and updated in GIS with geo-referenced coordinates for better identification, (3) it offers immediate availability of information on the newly constructed pipeline locations, and (4) it offers customer base information to the user groups to help them in related analysis, planning, and future projections such as new possibilities of pipelines, customers, gas volumes, and revenue including jobs to be undertaken by a third party. GIS may be used during the entire life cycle of the asset.

## 6.3 Built-In Design Features of the Natural Gas Distribution System for Operations and Maintenance

The natural gas pipeline distribution networks installed for transportation and distribution of gas to the various consumers must have the required built- in safety systems to ensure safe operation of the system.

1. **Design Features in the Spur Line**

   Spur lines are the pipelines that transport natural gas from the tap-off points of transmission lines to the city gate station (CGS). Natural gas distribution firms own the spur lines. These underground pipelines may be provided with three-layer polyethylene (PE) coating (or any appropriate coating) over it for protection against external corrosion. Apart from PE coating, cathodic protection (CP) is required also for these pipelines. A sectionalizing valve may be provided at appropriate locations. In case of heavy leakage of gas or fire, the gas supply in the affected section can be isolated by closing the related isolation valves. The valves may be operated both locally and from a master control room through the supervisory control and data acquisition (SCADA). Also, the pressure profile of the network at both the tap-off point and at the CGS is monitored continuously in the master control room through SCADA. This spur line must also have a facility for pigging.

2. **Design Features at City Gate Station**

   As mentioned previously, the CGS is a location where the gas supply to the city is initiated and monitored. All the process parameters such as pressure, temperature, flow and gas composition, and so on are monitored in the master control room building. The natural gas received at the CGS is at a high pressure (50–65 kg/cm$^2$), is filtered, and is passed through pressure control valve (PCV) and the shutdown valve (SDV) streams to reduce the pressure (to 15 kg/cm$^2$). PCVs help in maintaining the downstream pressure in the distribution network irrespective of the upstream pressure. SDVs are installed before PCVs which senses the distribution pressure. In case of any malfunctioning of the PCVs, the SDV stops the gas supply in the distribution network at a pre-set pressure that usually is kept slightly higher than the PCV set pressure. Two pressure safety valves (PSVs) also are installed in the line downstream of the PCV and SDV stream. These PSVs act as a third line of defense against uncontrolled increase in pressure in the natural gas distribution network. In case of failure of both (i.e., PCV and SDV), the natural gas distribution network pressure starts increasing at a set pressure, the PSVs come into operation and release the excess pressure. Finally, the gas passes through the flow meters where the flow of gas is measured before it goes into the distribution network. All the critical process parameters are monitored locally as well as in the master control room.

   Other available safety devices and equipment at the CGS are (1) LEL detectors installed around the process area, (2) fire water network, (3) break glass units, (4) fire extinguishers of various types, and (5) a dedicated communication system for communication with all pipeline installations on the network and with the master control room.

3. **Design Features in the Distribution Network**

   Primary protection to the steel pipeline against external corrosion may be given by PE/coal tar coating and cold tapes. Cathodic protection is planned also for the network apart from coating of the pipeline. Sectionalizing valves are provided in the network at regular intervals. A separate valve is provided also in each of the branch lines and at the inlet to each consumer point. In case of an emergency, gas supply of the affected section can be stopped by closing the upstream and downstream valves. Pipeline markers and warning signs are provided at a regular distance for easy identification of the pipeline. A continuous warning tape may be provided at the top of the entire natural gas distribution pipeline. A gas venting facility is provided at the CGS and also at each customer end for venting out all the gas from the respective natural gas distribution networks in case of any emergency.

4. **Design Features at Consumer End Terminals**

   Pressure reduction and metering skids are provided at each consumer end where the gas is filtered, pressure is reduced according to the requirement of consumers, and gas flow is metered before delivery to the consumer. These skids have the filter element, PCV, SDV, PSV, flow meter, and an electronic volume corrector (EVC) for proper regulation, flow measurement, and safety of the natural gas distribution system. In case of any emergency, an inlet valve is added in the pipeline before the metering skid is installed. It can be used for stoppage of gas supply.

## 6.4 Operations and Maintenance of the Natural Gas Distribution Network

The major operations activities are gas receipt, odorization, and pressure reduction (including heating the gas, if required). The other activities are operating and managing the district regulating station, field regulators, and gas metering for all categories of customers (i.e., domestic, commercial, and industrial).

O&M involves regular maintenance of all the assets of the natural gas distribution network and route patrolling to control third-party damages. It requires the natural gas utility to develop an offsite and onsite emergency plan and a disaster management plan to ensure safety of people and property along the pipeline route.

Some other O&M-related activities of the natural gas distribution business are (1) customer-focused operations such as correct metering and billing, pressure and flow requirements of consumers, shutdown, over draw, non-payment by the consumers, and consumer education and training, (2) preparedness for safety and emergency, (3) liaison with authorities and other utilities.

The gas measuring and billing including energy balance is the part of the operation. The safety, health, and environment including compliance of regulatory measures also are a part of the responsibility of the operating group. The maintenance activity covers the maintenance and upkeep of CGSs, District Regulating Stations (DRSs), field pressure regulating stations and end consumers facilities, sectionalizing valves, and other assets and facilities.

The O&M activities may be mainly classified as follows:

1. Corrosion control
2. O&M activities
3. Accidents and leaks

### 6.4.1 Corrosion Control

Corrosion control is required for all pipelines that are subject to corrosion. Corrosion assessment, monitoring, and mitigation programs must be developed by the natural gas distribution firm. These programs can be planned with inputs from operations, maintenance, and engineering. The character of the corrosion monitoring and mitigation program can be determined by many factors which may include pipeline product, environmental sensitivity and population density of surrounding area, pipeline operating history and age, operating pressure and operating conditions, and pipeline specifications.

Corrosion evaluation process is conducted through periodic review of operating conditions like pressure, temperature, and fluid analysis. If the analysis of operating conditions shows the possibilities of corrosion, then it is documented. The details of the analysis help in developing, monitoring, and mitigation program. The appropriate committee reviews the monitoring results and operating conditions in annual corrosion review meetings. If the rate of corrosion is found to have increased, then monitoring frequency may be increased or pipeline repairing is undertaken. This analysis is a continuous process to ensure the integrity of pipeline.

#### *6.4.1.1 External Corrosion Control*

The external corrosion of the pipeline is due to environmental conditions, flaws in a pipeline's external coating, inadequate cathodic protection, and physical contact that changes the pipeline or pipeline surface. External corrosion can be managed using a suitable external coating to protect the pipe. It is suggested (1) to have coating on all bare steel, (2) to insulate buried steel electrically from other conductors, (3) to protect the buried steel cathodically, (4) to provide for a means to test the performance of the system, and (5) to perform periodic tests and analysis.

#### *6.4.1.2 Internal Corrosion Control*

Internal corrosion takes place either if the dew point of the pipeline product exceeds the minimum operating temperature of the pipeline system or if the product contains fluids with free water, bacteria, oxygen, hydrogen sulfide, carbon dioxide, or suspended or dissolved solids that are corrosive in nature.

Internal corrosion mitigation can be planned through (1) installation of separation facilities to remove free water, or dehydration facilities to dry gas to bring its dew point below the minimum operating temperature of the pipeline, (2) addition of chemical treatment in batch form (with pig placement), in slug form (without pig placement), and/or in continuous injection form, (3) minimizing oxygen entering the system through vapor recovery units, chemical injections, routine pigging, and using inhibited methanol, (4) cleaning the pipeline using in-line cup-type pigs to remove solids and water, (5) removal of dissolved gases such as oxygen, (6) internal coating such as full-contact polymeric liners or free-standing non-corrosive liners, (7) use of inert pipeline materials such as plastic or fiberglass, and (8) bacterial treatment that can cause pitting using biocide programs.

The processes to manage external and internal corrosion may be developed which may involve processes related to monitoring of pipeline, atmospheric corrosion control, and monitoring remedial measures and corrosion control records.

#### *6.4.1.3 Records Related to Corrosion*

The natural gas distribution company also must maintain the following records and documents related to corrosion control: (1) cathodic protection design documents, (2) a soil resistivity survey report, (3) an electrical interference report, (4) inspection and maintenance reports, (5) material certification including dimension, metallurgy, performance, and functional report, (6) material test reports, and (7) approved drawings and documents.

### 6.4.2 Operations and Maintenance Activities

This section discusses O&M of equipment and assets, maintenance of the distribution network, and essential features of O&M plans and processes.

#### *6.4.2.1 Operations and Maintenance of Equipment and Assets*

The operations of a natural gas distribution business require managing all the major assets of the distribution network. Following is a discussion of the brief details about the functioning of the assets and its O&M:

1. **CGS and the Odorization System**: CGS receives high-pressure gas through spur lines and carries out filtration, pressure reduction, metering, and odorization before dispatching the natural gas to the steel distribution network.

   The assets to be operated at CGS are (1) isolation, vent, drain valves, (2) the knock out drum (KOD), (3) filter separators, (4) gas pressure streams: running and hot, stand by safety shut-off valve (SSV), and active and monitored safety release valve (SRV), (5) an odorant dosing system, and (6) pressure gauges.

   At the CGS, the activities undertaken are (1) monitoring of natural gas pressure, flow, and odorant level and smell, (2) odorant transfer from barrel to tank, (3) draining of condensate at the CGS using KOD, (4) leak check of all joints at the CGS, (5) testing of SRVs, (6) testing of pressure relief valve (PRV)/SSV, (7) housekeeping of the CGS facility, (8) calibration of pressure gauges, and (9) inspection of the odorant storage area.

In case of emergency, all the valves are operated for isolation, venting, and draining. The condensates also are drained in emergencies. Joints of KOD are checked for leaks. The shell and thickness of the KOD are visually inspected. All the filter elements at the CGS are cleaned and replaced to ensure their effective functioning. On-line testing and resetting of gas pressure stream may be conducted at a specific interval to ensure that the stand-by SSV and SRV are active.

The frequency of activities may be planned in line with the regulatory requirements and the company's own standards.

2. **Steel Distribution Network:** It carries natural gas at 15–19 bar from the CGS and takes it up to the common pressure regulating stations (CPRS), DRS, and compressed natural gas (CNG) stations. The assets under the steel distribution network are (1) coated steel pipeline, (2) crossings (road, railway, canal, and so on), (3) valve chambers, (4) isolation and vent valves, (5) a cathodic protection system, (6) a transfer rectifier (TR) unit, (7) pipe to soil potential (PSP) and test lead point (TLP), (8) anode ground bed (AGB), and (9) insulating joints.

   The activities to manage the network includes (1) patrolling and surveillance of the steel distribution network and crossings, (2) updating the networking drawings, (3) liquid penetrate test (LPT) leak detection/LPT survey of the steel distribution network, (4) preventive maintenance of the valve chamber (operation and greasing of valves and repair of valve chamber), (5) emergency patrolling and third-party coordination if work is going on or near the gas pipeline route, (6) monitoring of the TR (Transfer Rectifier) unit, (7) attending leaks and carrying out repairs for leaks, coating, and so on.

   The frequency for preventive maintenance of valve chambers, TR, AGB, and TLP may be according to the manufacturer's guideline. Preventive maintenance also may be required for the earthing pit.

3. **CPRS, DRS:** It receives medium pressure gas through the steel distribution line and carries out filtration, pressure reduction at 4–5 bar, and dispatch of natural gas to the MP PE (medium Pressure [MP] polyethylene [PE]) distribution network. The assets related to CPRS and DRS are (1) isolation, vent, and drain valves, (2) filter separators (F/S), (3) slam shut off valve, (4) active PRV, (5) monitor PRV, (6) SRV, and (7) pressure valve.

   The O&M activities at CPRS and DRS are (1) monitoring of pressure (inlet and outlet), (2) filter draining, (3) leak check of all joints, (4) Testing of creep relief valve, PRV, and SSV, (5) housekeeping in and around the fenced area, (6) calibration of pressure gauges, (7) attending leak, draining of condensates, and cleaning replacement of filter elements, and so on, and (8) general maintenance of the whole installation including painting.

Preventive maintenance includes filter element cleaning and replacement, on-line testing and resetting of the active monitor SSV and SRV, leak check, valve operation, paint touch up, and housekeeping, and so on. The O&M team also is responsible for leak repair, over pressure, under pressure, stoppage of supply (if required), and so on.

4. **MP PE Network:** It carries gas at 4–5 bar from the CPRS and DRS and takes it up to the service regulator (SR). The assets related to this network are (1) transition fitting, (2) pipeline, (3) valves with valve chamber, (4) route markers, (5) drawings, and so on.

   The O&M activities are (1) attending damages and leak complaints, (2) patrolling of PE network and crossings, (3) leak detection/LPT survey of the PE network, (4) preventive maintenance of the valve chamber, greasing and operation of the valve, and repair of the valve chamber, and so on.

5. **Service Regulators:** It receives medium-pressure gas through medium pressure (MP)polyethelene (PE) network and supplies the low-pressure gas to the low pressure (LP) PE network after filtration and pressure reduction. Related assets are (1) isolation, vent, and drain valves and (2) Over Pressure Shut Off (OPSO) , Under Pressure Shut Off (UPSO), and pressure gauges.

   The O&M activities are (1) pressure monitoring, (2) need-based leak check, (3) filter draining, cleaning, and housekeeping, (4) preventive maintenance of the SR mainly includes OPSO and UPSO, (5) on-line testing and resetting of the active monitor SSV and SRV, (6) leak check, valve operation, paint touch up, and housekeeping, (7) leak detection and arrest, over pressure, under pressure, stoppage of supply, if required, and (8) general maintenance of the whole installation including painting.

6. **LP PE Network:** It carries low-pressure gas at 900–1,200 millimeters per water column (mmwc) and supply to customer connections. This network has assets such as (1) transition fitting, (2) pipeline, (3) valve with valve chambers, (4) route markers, and (5) drawings.

   O&M activities are (1) patrolling, (2) leak survey and LPT valve chamber, (3) preventive maintenance for all assets, (4) leak detection and pipe damage repair and replacement, and (5) coordination with other departments.

7. **Domestic Customer Connection:** It consists of regulator, meter, GI/Cu pipeline and neoprene tube. Gas normally flows at 0.35 pounds per square inch (psi) and consumed at hot plate. Assets at domestic connection are (1) Transition Fitting (2) testing TEE, (3) isolation valve, (4) GI piping, (5) pressure regulator, (6) meter, (7) gas taps, and (8) rubber tubes.

O&M activities are (1) attending gas leak/gas smell complaints, (2) attending stove flame complaints, (3) attending disconnection requests, (4) attending reconnection requests, (5) attending customer request for alterations and modifications, (6) distribution of customer booklets and imparting awareness education to customers, (7) meter reading and billing, and (8) gas leak repair and replacement. Annual maintenance service is suggested for the whole installation from transition fitting to rubber tube. It may include low pressure gas test and leak detection test of the installation, checking and providing clamps and color touch up and checking the condition of the rubber tube.

## ILLUSTRATION: BILLING PROCESS

The company SYL has natural gas distribution business in three GAs and supplies natural gas as fuel to about 10,000 domestic household customers. The firm adapted a customer friendly initiative and opened local customer care centers (CCC) to address all customers' issues on supplies, services, complaints, safety, and so on at various centers (cities) where the company offers its services.

The issue of carrying out billing and payment-related activities for domestic household customers was becoming problematic for the firm due to large number of domestic household customers, each accounting for very low consumption of natural gas as compared to commercial, industrial, and transportation customers. The number of customers were growing faster month-by-month. Due to large number and fast-growing domestic household customer base, the billing process was becoming very time consuming and challenging. The utility was facing operational problems related to billing and payment activities for domestic customers. This problem was getting more aggravated as the domestic customers were spread out at different geographic locations in different districts. The cost of billing and payment activity was also increasing with the increasing number of domestic customers.

Problems Faced by the Company and Customers:

1. Meter readers had to visit every household, collect the meter reading, record the reading accurately on the meter reading sheet (chance of error while recording on the sheet), go back to office in evening, and hand over data to get it entered into computer software by a computer operator (here also there is a chance of error). At the end of a day, the meter reader could bring readings of only 30–40 customers. An additional staff

(*Continued*)

**ILLUSTRATION: BILLING PROCESS (Continued)**

(i.e., a computer operator) was required and was hired for data entry. This whole exercise of collecting meter readings took about 10–12 days, which kept on increasing due to the increasing number of piped natural gas (PNG) customers. The time taken to complete this meter reading activity was dependant on the number of persons hired as meter readers and the productivity of meter readers. As mentioned previously, one meter reader could at the most complete 30–40 household in a day. The complete billing process used to take about 23–35 days when the billing process was managed from the headquarters office and 20–29 days when the process was shifted to the CCC.

In the described process, the meter reader had to visit the customer's house twice to complete the billing cycle of a domestic customer; the first time to take the meter reading and the second time to deliver the bill. Current reading of domestic customers had to be keyed in twice to generate the bills; the first time in the printed sheet (hand written readings) at the time of meter reading and the second time for keying in to the Excel spreadsheet to upload in the software.

2. While carrying out the described exercise, if the customers find errors in the billing, they contacted the local CCC offices. CCC executives receive many calls related to customer complaints which caused dissatisfaction and distrust among the PNG customers.
3. Customers were generally given 6–7 days for payments. To make payment, customers had to use their time to travel and visit the CCC and deposit the payments.
4. After the payment is made, the personal computer operators must make entries again in the software and issue receipts. The accounts and finance department had to deposit the payments to the bank daily, monitor the check bouncing cases, carry out payment default and delay analysis, reconciliation of accounts, and so on. All this was becoming a difficult task for the company and it was found that almost always, throughout the day, the CCC executives were busy doing these activities.

**SOLUTION TO PROBLEM**

In the light of the described problems, the firm decided to improve its PNG billing system by improvising and redesigning the internal processes

(*Continued*)

**ILLUSTRATION: BILLING PROCESS (Continued)**

through techniques of process analysis. The firm took up a challenge to redesign the entire process of the PNG billing system to make it smooth, easy, less time-consuming, convenient, and facilitating for the customers. It was also aimed at optimizing the manpower resources of the organization. The management of the firm formed a committee of executives from business development, O&M, accounts and finance, and the information technology (IT) departments to find the solution. The Chief Executive Officer (CEO) of the firm was also a part of this committee. The committee initially looked for a solution from the practices followed by other peer gas utility companies.

1. The committee found that some companies had completely outsourced these activities and were paying USD 0.10 per bill, on an average, to the contractor. The studied firm did not want to follow this system and ruled out this idea for following reasons:
    - The firm had set up local CCC office to be closer to the customers so that customers' complaints and problems can be dealt with directly. This was done to make the customers' experience with the company's services in a personalized manner.
    - The Company wanted its personnel to visit each domestic PNG customers and take a look at the safety and operations aspects, check leakages and system safety etc. along with meter reading and billing. Additionally the company also had an opportunity to hear complaints, if any, from the customer, which otherwise the customer might not have communicated.
2. The committee also found that some companies follow the practice of average billing every month, depending on the average consumptions of individual customer. Such firms finally reconcile the billing with the exact meter reading taken annually. The company was not very keen to follow this system due to same reasons as explained previously in (1).

**CREATIVE AND JUSTIFIABLE SOLUTION IDENTIFIED BY THE FIRM**

The goal of the firm was to eliminate some of the non-value added and duplicate activities of the billing process and at the same time maintain the personalized touch with the customers. After detailed deliberations

*(Continued)*

**ILLUSTRATION: BILLING PROCESS (Continued)**

**FIGURE 6.1**
Handheld device for billing.

among the committee members, they considered billing with a handheld billing device (see Figure 6.1). They got this idea from another public utility system (i.e., the state road transport department) wherein the bus conductors use such devices. The idea seemed remotely possible. The firm observed the following points while evaluating this solution:

- The handheld billing device must be compatible with the billing software at local CCC office and headquarters office
- The device must be compact, easy to operate, and weather proof
- The device must not be very costly
- The device must be able to handle enough quantity of data of customers who can be covered in a day

With these points in mind, the company started exploring markets to find vendors and suppliers of such products. They also contacted the supplier who provided billing software to the company. The company thought that the billing software supplier would be the best agency to help them in this matter. The software supplier did agree; however, the supplier quoted a very high price (i.e., USD 3,000) for supply of the handheld billing devices. They also assured the company that the device would be compatible with the existing billing software.

*(Continued)*

**ILLUSTRATION: BILLING PROCESS (Continued)**

The firm found the quoted price too high for just one handheld device and dropped discussions with the supplier. After several efforts, the firm found another supplier of handheld devices. But the supplier's constrain was that their device was not compatible with the company's billing software. The company promised the supplier to extend all assistance, cooperation, and help to develop compatibility by a universal serial bus (USB) port, which could then be connected with the cable to computer and data interface could take place. After many discussions, many rounds of meetings, and deliberations, the device could be developed with software compatibility. After all the meetings and negotiations, the efforts paid off and the supplier agreed to supply the device at a cost of USD 400 per device. All the processes of meter reading, billing, payment, and so on were redefined after the handheld billing device was finalized. The output of meter reader went up by 100 percent and now the meter reader could visit 75–80 customers daily.

Benefits to customers with the improved billing system through a handheld device:

1. Instant billing in front of the customers' eyes provides utmost satisfaction and trust to the customer.
2. Bills generated immediately, prints taken, and delivered to the customers on the spot.
3. Minimizes the chances of error. Errors if any can be corrected immediately on the spot.
4. Customers can choose to make on the spot payments by check.
5. Bill payments became faster; many customers chose to pay immediately on the spot so that the visit to local CCC office could be avoided. The firm is now planning to make an improvement in handheld devices that can immediately generate and print payment receipt.
6. The new billing system helped the firm increase customer satisfaction and develop trust and goodwill among customers.

Benefits to organization:

1. The device is compact, weather proof, easy to operate and can be used by any unskilled person.
2. Meter reader can now cover 75–80 customers daily as compared to 30–40 daily using the old billing system.

(*Continued*)

**ILLUSTRATION: BILLING PROCESS (Continued)**

3. The number of days in billing cycle reduced from 23–35 days to 13–19 days.
4. On the spot payments by customers through check further improved the firm's revenue flow.
5. At the end of the day's billing, customer's data in handheld billing device is downloaded using the USB port and updated in office software data. The daily updating of data also eliminated account reconciliation problems.
6. The firm's purpose of providing personalized services was fulfilled. While visiting domestic customer's premises, company personnel can also check for any leakages. Safety-related issues and complaints, if any, can be immediately addressed.
7. Huge savings in terms of the time devoted by employees in the billing activity, which led to a savings in manpower and operational cost. The CCC executives and personal computer operators saved lot of time and energy, which could be utilized for many other services.

(*Source*: Yadav, S., and P. Paliwal, "Re-Engineering service delivery process: Case of a natural gas utility," *Journal of Services Research* (*JSR*), 11(2), 2011–2012.)

8. **Commercial Customer Connection:** It consists of regulator, meter, GI/Cu pipeline and neoprene tube. Gas for commercial connection flow generally at 250 mmwc and consumed at hot plate. O&M activities are (1) annual maintenance service which includes leak test, rubber tube replacement, clamping, paint touch up and (2) attending to the complaint of gas smell, leak, fire, explosion, over pressure, under pressure, no supply of gas, and so on.
9. **Industrial Customer Connection:** It consists of filter, regulator, meter, PE and carbon steel pipeline; gas flows at 1.75 psi and is consumed at gas equipment such as a boiler, furnace, and so on. It also will have similar assets as in a domestic connection. O&M activities are (1) monitoring of pressure, flow, and volume daily, (2) attending complaints, and (3) annual maintenance service.

10. **Control Room, Office, and Store Premises:** O&M personal at the control room carry out round the clock (24 hours; 365 days) operational activities for the natural gas distribution utility. Stores stock the inventory for O&M.
11. **Others, Utilities, and Supporting Assets:** (a) Portable gas detectors must detect gas presence during leaks, escape, and maintenance jobs, (b) maintenance vans are required for attending regular and emergency activities, (c) tools and tackles for PE repair and maintenance are required for repair and maintenance of leaking, damaged PE pipes, and fittings on the PE network, (d) tools and tackles for CGS and SR are required for repair and maintenance and online functional testing of valves for the CGS and SRs, (e) tools and tackles for repair and maintenance of customer connections are for repair and maintenance of customer connection, (f) tools and tackles for odorant handling are for storage, transportation, transfer, and waste disposal; they are required for transfer of odorant from barrel to storage tanks and neutralizing the residual odorant and vapor, (g) PPE are for personal protection during normal and critical jobs, (h) fire extinguishers for extinguishing accidental fire and during routine jobs, (i) signage, caution, and information boards for display of information and caution, (j) mobile phones for quick and effective transmission of information, even when away from control room, (k) public announcement system for making public announcement to customers during gas supply stoppage and resumption and festival safety, (k) wireless sets for quick and effective transmission of information, even away from control room, (l) computer system and printers for billing, records, reports, and so on, (m) log books and stationeries for logging daily information and parameters, customer complaints, handling and status of complaints as per shift, including attendees in shift.

    Furthermore, (1) all the tools and tackles need regular inspection and maintenance, (2) the maintenance van must be checked daily and if required emergency repair maintenance is carried out, (3) portable gas detectors require calibration/repairing, (4) fire extinguishers require inspection and refill after use, and (5) wireless sets require inspection, servicing, repair and maintenance.
12. **Calibration, Testing, Inspection, and Audit:** O&M activities also include (1) calibration of pressure gauges, (2) calibration and testing of gas detectors, (3) inspection and audit of CGS, CPRS, and DRS, (4) inspection and audit of SR, (5) inspection and audit of domestic, commercial, and industrial connections and (6) inspection and audit of the CP system.

## 6.5 Maintenance Processes

As known, the safest and most cost-effective form of maintenance is preventative maintenance. An effective corrosion control program and a well-coordinated inspection routine are necessary to minimize repairs and ideally eliminate hazardous conditions. Preventative maintenance programs are developed to assist the O&M staff to maintain the natural gas distribution network. O&M personnel must know the location and surface access to installations such as valves, vents, and cathodic protection rectifiers for pipelines and other system equipment. Some of the activities related to maintenance of network are discussed in the next section.

### 6.5.1 Patrolling and Surveillance

The patrolling schedule must be such that the entire primary network and secondary network is inspected at regular intervals to observe surface conditions, construction activity, encroachments, soil wash outs, and any other factors that may affect the safety and operation of the network. It also includes a check for area population, development change, pipeline marker condition, test station markers, and so on. All exposed piping must be inspected for coating deficiencies and/or other damage. The frequency of patrols may be planned appropriately considering the regulations and the company's own standards.

### 6.5.2 Annual Leakage Survey

A leakage survey of all pipelines must be performed at regular intervals according to regulatory requirements and company standards. The survey is undertaken using a combustible gas indicator, flame ionization leak detector, or infra-red unit. After completion of the survey, the results are recorded and reported as required by regulations and company standards. Leakage surveys using gas detectors must be done in accordance with the requirements of ASME B 31.8. Gas detectors, duly calibrated, must be available at all times in ready use conditions for emergency surveys and use.

### 6.5.3 Pipeline Marker and Signage

Signages are erected along the pipeline and they are inspected periodically. Signage must be structurally sound and legible. It must fulfill regulator criteria. If the pipeline is removed, then signage is also removed. Pipeline warning signs are required where pipelines cross highways, roads, railways, or water courses. Pipeline warning signs must be placed in areas adjacent to all pipeline installations including meter stations, valve sites, risers, and line heaters. A large facility identification sign is required at the entrance of any licensed gas compressor station. The sign must include the facility name, legal location, licensee, emergency telephone number, and the appropriate warning symbol. When a pipeline or portion of a pipeline is removed, O&M personnel must remove related signs.

Markers may not be installed for service pipeline within consumer premises; however, the operating company must maintain such service pipeline routing drawings for easy reference. The operating company must provide minimum safety information to the consumer before starting the gas supply.

### 6.5.4 Repairing Pipelines

To initiate the repair work, first the pipeline ID, status, and substance must be identified. Health, safety and environment ground disturbance procedures must be followed before replacing any portion of the pipeline. After the repairing is done, the required report is completed. Pressure test is conducted for the repaired pipeline segment and the test report is submitted to the regulator. The pipeline is put back into service after the regulator's approval.

### 6.5.5 Abandonment or Inactivation of Facilities

The O&M team must update the master spreadsheet when a pipeline's status changes and must review the spreadsheet at regular intervals. The steps taken when abandoning pipelines are (1) leave the place in safe condition, (2) disconnect all sources of gas, (3) cut and cap below ground level, except when within a facility that will continue to operate, in such cases just disconnect and cap, (4) remove cathodic protection, (5) remove surface equipment such as line heaters, pig traps, and risers, unless located within an operating facility, and (6) leave no stagnant fluid traps. Remove underground T-connections and replace them with straight pipe. Abandoning, disconnecting, or reinstalling distribution facilities must be according to ASME B31.8. Any activity associated with abandoning, disconnecting, or reinstalling of distribution facilities must require a work permit issued by the authorized person.

### 6.5.6 Compressor Station Maintenance

Appropriate starting, operating, and shut down procedures must be followed for the maintenance of compressor stations. Inspection and testing of pressure relief devices are undertaken at regular intervals as specified by the regulatory authority. Each pressure device must be inspected for (1) capacity, (2) proper set pressure, (3) visible damage, (4) mechanical condition, (5) protection from dirt, liquids, or freezing, and (6) proper installation and any condition that could cause the device to malfunction. Compressor stations are equipped with a gas detection and alarm system.

### 6.5.7 Valves and Pressure Limiting Devices

These components are checked periodically. The reliability and compatibility of the components is checked. The defective components are repaired. After repair the function test is carried out and records are updated. Pressure control and overpressure protection devices ensure the pipeline operates below its MAOP. Pressure devices are grouped into (1) process control devices, (2) pressure limiting devices, and (3) pressure relieving devices.

### 6.5.8 Valve Maintenance

All pipeline isolation block valves that may be used in an emergency are inspected at regular intervals. The actions that are performed and documented on the valve inspection are (1) confirm the presence and relative position of the valve on the system map, (2) confirm the presence and condition of the line marker, if required, and the legibility of the writing, especially the emergency telephone number, (3) visually inspect the valve and piping for corrosion and any indication of acts of vandalism, (4) inspect the valve and piping for external leakage using either a combustible gas indicators or a soap solution, (5) function test and partially operate the valve to assure that the stem has not seized up, and (6) for remotely controlled valves, although the device can be triggered from an offsite location, an observer must be onsite to record that the valve physically moves.

### 6.5.9 Prevention of Accidental Ignition

Site specific risk assessment must be carried out before commencing any repair activities. The outcome of such risk assessment must be documented and considered while preparing a safety plan for the repair work.

The steps used to minimize the danger of accidental ignition of gas in any structure or area where the presence of gas constitutes a hazard of fire or explosion are (1) assess if the work may be moved out of the hazardous location, eliminating it as "Hot Work," (2) ensure proper isolation of all potential sources of energy or combustible substances, and (3) utilize portable gas detectors; capable of detecting at minimum 02 and LEL; check thoroughly to determine the presence of a combustible gas mixture prior to welding in or around a structure or area containing gas facilities; do not use gas or electric welding or cutting on pipe or on pipe components that contain a combustible mixture of gas and air in the work area; begin welding only when safe conditions are validated and continuously monitor conditions during activities for presence of combustible gases, (4) provide a fire extinguisher and/or fire watch, (5) maintain site control of the area limiting access; post warning signs where appropriate, (6) when flashlights, hand lanterns, or other electrical equipment are needed for a task, intrinsically safe (Class I) devices must be used, and (7) install a metallic bond at the location of a cut in a pipe that is made by other means than with a cutting torch.

### 6.5.10 Plastic Pipe Maintenance

The safety precautions which must be ensured during emergency repairs or breakdown maintenance of plastic pipelines are (1) all naked flames and sources of ignition must not be allowed in the immediate work area; (2) gas level must be monitored during the repair work with gas detectors; the repair must not be carried out in an atmosphere which contains natural gas; and (3) adequate fire extinguishing equipment must be available during such repair.

Squeezing-off and reopening a thermoplastic pipe or tubing for pressure control and repair of plastic pipe or tubing must be in accordance with ASME B 31.8.

### 6.5.11 Miscellaneous Facilities Maintenance

Flexible steel braided hose used to connect consumer appliances must be inspected at least once every year for leakage, kinking, corrosion, abrasion, or any other signs of wear and damage. Any hose worn out or damaged must be removed from service and replaced.

#### ILLUSTRATION: STUDY OF MAINTENANCE OF RISERS IN THE DUCT IN HIGH-RISE BUILDINGS

The natural gas distribution company installed the above ground supply pipes (risers) in the duct systems of high-rise buildings within the city area. A survey of risers installed in ducts in the high-rise buildings was conducted for maintenance of the domestic connections. About 16,000 risers were physically surveyed to identify the maintenance requirements. The survey involved study of various processes like riser installation, annual maintenance requirements, replacement of corroded parts, and so on. Possibilities of using alternate or other material instead of GI supply pipes also were considered.

Study Findings revealed that (1) many times the risers were installed near drainage pipelines; in case of any leakages in the drainage line, the natural gas GI pipes get corroded; installation of risers in the ducts must be avoided at all times, (2) in cases where the pipes were installed through the walls, it got corroded as the pipe came in contact with *reinforced cement concrete* and; whenever the pipes are installed through the walls, it must be wrapped with high-density polyethylene (HDPE) sleeves where it touches the wall, (3) electrical wires were wrapped around riser pipes; customer awareness is required to avoid wrapping electric wires or any other hazardous material on the risers, (4) the riser must be installed up to the final floor in a high-rise building irrespective of the number of connections at the time of installation; this will help in future for connecting the new potential customers, (5) the annual inspection must be done as required by the regulations and wherever any deformity is found in risers it must be resolved immediately, (6) zinc coating may be applied instead of powder coating on GI pipes, (7) Cu tubing for the above ground supply pipes (risers) may be a better replacement compared to the GI pipes (Figures 6.2 through 6.5).

(*Continued*)

**ILLUSTRATION: STUDY OF MAINTENANCE OF RISERS IN THE DUCT IN HIGH-RISE BUILDINGS (Continued)**

**FIGURE 6.2**
Electrical wire wrapped with the riser.

**FIGURE 6.3**
Riser with drainage pipes in ducts.

(*Continued*)

**ILLUSTRATION: STUDY OF MAINTENANCE OF RISERS IN THE DUCT IN HIGH-RISE BUILDINGS (Continued)**

**FIGURE 6.4**
Riser passing through wall.

**FIGURE 6.5**
Risers passing through wall.

### 6.5.12 Statutory Compliance Monitoring

All the statutory compliances of concerned country government and regulatory authorities must be handled. The statutory compliances may be with respect to (1) the Factory Act Requirements and the factory license, (2) SRV and pressure vessel testing, (3) stability certificate of civil structure of company premises and plant building, (4) pollution control related compliances, (5) power-related compliances of the concerned authority, (6) the Vehicle Act requirements for emergency vans, (7) insurance (i.e., third-party public liability insurance, manpower insurance, asset insurance), (8) reports (i.e., daily activity report, monthly report (review for current month and plan for next month), gas smell monitoring and complaint analysis report, TR monitoring report, CPRS and DRS monitoring report, monthly reconnection and disconnection (RC and DC) report, network health and integrity report, shut down report, break down report, activity report, awareness and training report, near miss report, hazard report, accident report, major gas leakage, fire, and explosion report, and so on.

## 6.6 Essential Features of the Operations and Maintenance Plan and Processes Related to Operations and Maintenance

Some of the essential features of O&M plans and processes to operate and maintain the natural gas distribution system follows.

### 6.6.1 Operator Qualification and Training Program

A newly recruited field operator is mentored by the qualified operator or foremen. The mentor observes the newly recruited field operator complete critical tasks. Thus, a newly recruited field operator receives orientation. If the tasks are adequately completed, it is assumed that the newly recruited field operator can work unsupervised.

Industry and site-specific training, along with competency reviews, are tools used to train new field operators. Monthly safety meetings also are used as a forum for learning wherein safety topics and pipeline incidents may be discussed.

The training program must cover (1) the hazardous characteristics of gas, (2) familiarization with commissioning, operation, and maintenance procedures, hands on experience on operation of emergency and manual shut down systems, (3) effective isolation of any gas leak, (4) safety regulations and accident prevention, (5) firefighting equipment operation and its upkeep, and (6) first aid and housekeeping.

The training process must be subjected to periodic internal audits to ensure effective implementation and improvement. Training must include mock safety drills at least twice a year. Training program also must envision imparting training to employees and contractors of other utility companies sharing the same corridor to make them aware about hazards associated with leak and damages.

### 6.6.2 Identify and Document All Covered Tasks for Operators

Tasks are operations-based (sending and receiving pigs, handling chemicals, gauging tanks, changing orifice plates, and so on) with a specific focus on safety and loss control. Depending on equipment in the production area, operations tasks cover not only pipelines, but also other facilities like CGS, DPRS, IPRS, SVs, and so on. These tasks are discussed in a previous section about O&M of equipment and assets. Processes are established at each field office to periodically review and assess specific operating procedures.

### 6.6.3 Third-Party Companies and Contractors Responsibilities

In case the work is entrusted to a third party, it is ensured that the third-party company personnel are qualified and able to do the work.

### 6.6.4 Pipeline Start Up and Shut Down Procedures

Pipeline start up and shut down procedures are critical tasks. All appropriate precautions must be taken to ensure maximum operating temperatures and pressures are not exceeded for even a short period of time. It is ensured that valves are in the correct position and are opened or closed in the proper sequence, pumps or compressors are started and shut down safely, and safety shut downs are properly calibrated and function tested according to the required maintenance schedule.

### 6.6.5 Class Location Review

These activities include the assessment of actual area affected by increase in the population density, any physical barriers, or other factors that may limit further expansion in the area. Data and observations supporting the need for a class location review are recorded during the pipeline patrolling. If significant changes in these parameters are found, then the pipeline segment class location is changed.

If the study based on class location review reveals that the hoop stress corresponding to the established MAOP is not acceptable for the new class location, either the MAOP must be lowered or other appropriate action must be taken.

### 6.6.6 Damage Prevention Program

The damage prevention program must consider efforts to minimize the possibility of damage to the pipeline by third-party excavation. A proper procedure may be laid down for safe digging. Company representatives must remain present on site at the time of excavation. When a natural gas utility performs any type of survey of the pipeline, or an employee is traveling in the area of the pipeline for any reason, that person must make a conscious effort to observe activity or signs of activity related to excavation in close proximity to the pipeline right of way (RoW). If an employee observes any excavation activity that is believed to be encroaching upon the RoW, or has the potential to execute upon the RoW, the employee must advise the excavator to stop the observed activity immediately.

### 6.6.7 Blasting

Prior to any blasting associated with excavation in the nearby area of a pipeline, O&M personnel must secure an engineering opinion of the potential stresses imposed upon the pipeline in consultation with integrity engineering. Inspections of the pipeline must be performed prior to and after blasting activities to verify the integrity of the pipeline. Each inspection must include a leakage survey. No blasting may be carried out within city limits and near any third-party structures or facilities. As such, blasting must only be used after proper authorization from concerned authorities even if it is safe to carry out such operations.

### 6.6.8 Emergency Response and Disaster Management Plan

The major hazard in gas supply is a leak or rupture of a pipeline, which results in an uncontrolled gas release. The gas leaks may prove hazardous if not handled properly and disposed of. Therefore, safe, timely, and organized actions are essential.

Possible causes of gas leaks are (1) pipeline damage due to construction or material failure, corrosion, or mechanical damage, and so on, (2) flange leak due to material failure, wrong gasket (inside the customer's premises only), (3) valve leak due to stem seal failure, assembly failure, and so on; valve seal leak can occur as a result of ageing, filthiness, damage, wrong gasket used, or assembly failure, and (4) enemy action, sabotage, or natural calamities such as earthquake, flood, and so on.

The ERP helps to minimize hazards which may arise from or be associated with gas pipeline emergencies. Through a well-designed ERP, the utility (1) receives notification of emergencies involving its pipeline system, (2) efficiently accomplishes required communication of the occurrence, (3) has immediate access to established written procedures to respond to foreseeable emergencies, (4) coordinates with local fire and police agencies before, during, and after the occurrence of an emergency, and (5) pursues appropriate follow up activities after an emergency. The ERP enables response promptly and effectively to different types of emergencies such as (1) gas detected inside or near a building, (2) fire located near or in the pipeline facility, (3) explosions occurring near or in the pipeline facility, and (4) natural disasters.

The emergency control room is operated by the utility and is manned around the clock and equipped with effective communication system and emergency vehicles fitted with communication facilities, first aid equipment, fire extinguishers, gas detectors, repair kits, and tools, maps, plans, material safety data sheets, and so on at its disposal.

While preparing the ERP and disaster management plan, the entity must take into confidence the various local authorities (i.e., the fire authorities, police authorities, health authorities, local administration, disaster management authorities, and so on) and clearly elaborate on their role in case of an incident.

Written emergency procedures are laid down which may include (1) "do's and don'ts" during emergency and other safety instructions, (2) telephone numbers of emergency response team members, emergency services, concern authorities, law enforcing agencies, contractors and vendors, fire services, and so on, (3) actions to take during an emergency including warning and cordoning off the affected area and informing the concern authorities and/or other utility companies affected by the emergency.

In case a significant fire develops, the personnel will not be able to approach the point of leakage or extinguish the fire because of the pressure of the gas escaping and the intense heat radiated by the fire. Even if it is possible, immediate extinguishing of the flames may be dangerous because there would be a "flash-back" as the leaking gas may catch fire from surfaces heated by the fire. The procedure for tackling such incidents is (1) ensure the whereabouts and safety of all personnel, (2) inform the fire brigade and safety department, (3) isolate the section of the pipeline from which gas is leaking, (4) protect the section of the pipeline near where the gas is leaking, (5) only when the fire has been greatly reduced by loss of pressure and is under control may it be extinguished by power appliances. Water hoses and monitors must continue to cool surrounding equipment until the gas leakage stops. If a gas leakage cannot be shut-off completely, it would be safe to leave the fire burning in a contained manner until positive shut-off is established.

The plan is prepared in consultation with all the agencies and departments involved in handing of an emergency which arises because of a gas leakage or a fire. Copies of the plan are made available along with the gas network maps to all the concerned authorities and executives. Key features of the plan may include (1) description and details of the gas pipeline network, (2) identification and definition of roles, responsibilities, and authorities, (3) procedures for emergency handling, shutdowns, and evacuations, (4) emergency contact numbers, (5) list and location of emergency tools and spares, and (6) training requirements and plans.

**ILLUSTRATION**

One of the incidences that took place in the city gas network is presented below to highlight the preparedness of firm to handle an emergency.

Incidence: Gas leakage and fire in the 3″ branch line valve

Place: 3″ tap off line near the national highway

Date and Time: 8.20 PM, 6.4.2002 (Saturday)

After a shutdown of 7 days (from March 30, 2002 to April 6, 2002) for the city gas network expansion works, the gas supply to Firozabad consumers was resumed on 7:30 pm on April 6, 2002. Soon after the resumption of gas supply at 8:20 pm, a phone call was received from one of the consumer informing that there was a fire on the gas pipeline near his factory in the 3″ tap-off line connection. The executive who was available at the CGS at that time responded immediately by shutting off the gas supply to the city network and started venting off the gas from the CGS. In about 20 minutes, the entire network was depressurized and the fire was extinguished on site. There was a quick response from the fire department and local administration. Fire tender reached the site quickly and along with police they cordoned off the entire area and kept the surrounding area cool to prevent the fire from spreading. The fire was extinguished after the gas supply was stopped from the CGS and the network was depressurized. Thereafter, the excavation was done and further action was taken to rectify the problem and restore the supplies. The total system was restored and gas supply was resumed to consumers by 12:30 on April, 7, 2002.

**LOSS AND DAMAGES**

No casualty occurred due to the leakage and fire. Also, no significant damage occurred except for damage to belongings (line clothes, bed, food grains, utensils, and so on) of a guard living in the adjacent premises.

**REASON AND CAUSES**

The 3″ valve was installed in the underground branch line during the present shut down and after installation it was covered with sand bags and loose soil so that no one would fall in the valve pit while the valve chamber was being completed. It was suspected that some heavy vehicle crossed over the pit and caused rupturing of the valve body. Rupture in the heat affected zone portion of welding in the valve body

(*Continued*)

**ILLUSTRATION (Continued)**

resulted in the leakage as soon as the gas pressure was increased in the network after resumption of shut down. A spark from the overhead electrical lines provided the source of ignition.

**LESSONS LEARNED FROM THE INCIDENCE**

The case though a very small one which didn't result in any loss of life or property provides a deep insight into the preparedness of any organization in responding to an emergency. It is not always necessary to wait for a casualty to happen before taking corrective actions to strengthen the safety systems. This case highlights the importance of the key elements of an emergency response and disaster management plan and why proper coordination between all these elements is essential to effectively handling any kind of an emergency in the network. Timely action from all the agencies involved and their knowledge and skills helped in preventing any major loss and casualty in the present case.

(*Source*: Reproduced from the paper by Ayush Gupta, Preparedness to handle emergency in city gas distribution networks, *4th Pipeline Technology Conference*, 2009.)

### 6.6.9 Public Awareness Program

The natural gas utility must develop and implement a written continuing public education program that follows the guidance provided by the regulations. The goal of the public awareness program is to promote public awareness (consumers and general public), which improves public safety and protects the assets and environment. The educational material may be prepared in the local language. Local audiovisual media available must be used for such educational programmes.

### 6.6.10 Pressure Testing Pipelines and Pressure Limiting Devices

After the pipeline is installed, pressure test is conducted. The medium of the test may be water and accordingly the regulatory authority is notified for testing. The approval of the regulator is required to conduct the test. In case of failure in the test, the regulator is informed and repairs are undertaken. In case of success, the test results are documented. Pipelines must be tested for strength and leaks by replicating the same conditions under which the pipeline will operate. A segment of pipeline that has been reconstructed, relocated, replaced, or reactivated may not be operated until it has been tested. Any leak that is located must be corrected prior to placing the pipeline in service.

Due consideration and respect for the pressures while testing at elevated pressures is required to ensure safety and environment protection. All equipment including hoses, fittings, pumps, and so on must be rated for the pressures required at test and they must be in good condition. Hydrostatic test water must be disposed of in a manner sensitive to the environment as required by the regulations.

The capacity of all pressure-limiting devices must be checked at least once each calendar year not to exceed an interval of 15 months to assure their ability to limit the maximum allowable operating pressure of the pipeline.

### 6.6.11 Abnormal Operating Conditions

O&M personnel must take appropriate steps to investigate and correct the abnormal operating conditions (AOCs) so that the pipeline returns to normal operating conditions. Some of the AOCs are (1) unintended or strange valve movement, closure, or opening, (2) unintended or unexplainable pump, compressor, or facility shut down or start up, (3) unscheduled abnormal variation in pressure, flow rate, or temperature, (4) loss of critical communications, (5) unintended, strange, or improper operation of a safety device (including pressure limiting or shut down devices), and (6) other incomprehensible malfunctions of components, deviations from established operating limits, or personnel error that could become a hazard to personnel safety or result in impacts to the environment if allowed to continue.

The O&M personnel must document the occurrence, investigation, response procedures, and follow up activities of the AOC in a report. The report may also include recommendations to eliminate the cause of the AOC.

### 6.6.12 Effective Liaison with Concerned Authorities and Other Utility Departments

From experience, the incidents and accidents on gas pipeline networks worldwide are because of third-party damages. Therefore, the natural gas distribution company must have a close coordination and liaising with the concerned authorities and utility departments like telecommunications, city municipal departments, water works, public works department, electricity authorities, and so on. Any development activity by any of the departments in the vicinity of pipeline network may cause an accidental damage to the pipeline and vice-versa. It is also suggested that while laying the pipeline, care must be taken by other utilities coming close to the pipeline route. The "One Call System" may be planned and implemented by the authorities through regulations that can bring all the utility departments onto one platform.

### 6.6.13 Pipeline Failure Investigation

Besides reporting and recording of all instances of asset related failures, damage to the environment and third-party property must also be recorded. The failure investigating team must be comprised of personnel trained in failure investigations. The data from all failure occurrences must be analyzed for trends so that appropriate actions including training could be taken to minimize failures.

### 6.6.14 Records

Besides the details of leak records as covered under ASME B31.8, the natural gas distribution company also must maintain the following records and documents: Design specification, alignment sheets for primary network of steel pipeline and associated pipeline network, other installation and test records, surveillance inspection and maintenance records, material certification including dimension, metallurgy, destructive and non-destructive testing records, performance and functional test reports, welding records including procedure qualification record, welding procedure specification, and welder qualification records, commissioning reports, non-conformance and deviation records, calibration records of inspection, measuring, metering and test equipment, audit compliance reports, statutory clearances, approved drawings and documents, HAZOP/risk assessment reports and compliance to recommendations of such reports, all operation and maintenance manuals, and so on.

### 6.6.15 Other Operations and Maintenance Processes

O&M procedures must also address the following: (1) preventive maintenance plan and procedures in accordance with recommendations of the original equipment manufacturer(s), (2) a well-designed system of periodic inspection for all facilities, (3) a calibration plan for meters, gauges, and other instruments affecting quality and safety of system, (4) a plan for functional testing of pressure regulation and control equipment (active/monitor regulator, slam shut valve, pressure relief valves, control valves, and so on), (5) isolation scheme (complete with drawings) showing the orientation of the facilities, location of major services, power switches, entry and emergency exits, fire assembly points, and so on, (6) main components, including their identification number, (7) limits of operating parameters (pressure, temperature, flow, levels, and so on), (8) work permit procedures to be followed by maintenance personnel for protection of property from damage and fire, and so on, (9) procedures to log operation and maintenance activities, and (10) "do's and don'ts" and safety precautions during operation and maintenance.

## 6.7 Accidents and Leaks

In case of pressure decrease, product volume imbalance, or report from public about leak or rupture, the offending pipeline segment is shut down and the regulator is notified immediately. In case of any internal or external pipeline failure, the repair is completed and appropriate report is submitted to the authorities. Otherwise, the scheduled inspection is undertaken and completed to check for accidents and leaks.

1. **Report of Accidents:** All incidents involving natural gas distribution facilities that result in injury or death to any person, damage to property, and so on must be reported immediately by telephone to the concerned authorities. According to the regulations, the written report is submitted to the authorities giving a detailed description of accident (incident), response, action, and investigation by the operator. Results and findings of the investigations also are reported.
2. **Leaks and Ruptures:** Leaks can be small openings, cracks, or holes in the pipeline that cause a slow release of pipeline product. They may be seen or smelled during RoW inspections, or they may be indicated by slight variations in pressure meters or product volume imbalances. Ruptures are instantaneous tears or breaks in the pipeline that release an immediate and large volume of pipeline product. They can be detected quickly because of the significant reduction in operating pressure and product volume imbalance. A pipeline leak or rupture is considered a pipeline failure. A pipeline hit can be caused by contact from a backhoe, shovel, hydro-vac unit, or other third-party event. Steps to avoid pipeline hits may be planned and implemented.
3. **Leaks: Records:** Records of the entire history of a leak from discovery through repair to follow up inspection must be maintained and each leak may be identified by number. The name of the responsible individual also must be on the records.
4. **Leaks: Instrument Calibration:** All the instruments utilized for leakage surveys and investigations must be tested against a known sample or in accordance with the manufacturer's recommended instructions. It must be tested after any repair or replacement of parts other than normal maintenance if required instruments are calibrated.
5. **Leaks: Response:** The ERP may be followed by field operations when necessary. All leaks found on the pipeline system must be treated and handled as early as possible. If required, production may be stopped. In case of leaks, the nearest valve on each side of the leak may be closed to block in and isolate the pipeline segment involved. This blocked

pipe segment will then be depressured and the leak excavated. The leak will be repaired and the pipeline is placed back in service.

6. **Accident and Emergency Reporting Procedures:** The entity must put in place a documented in-house accident reporting procedure and its response plan for all kinds of accidents and emergencies such as (1) near miss accidents, (2) accidents without loss of production, supply, or human life, (3) accidents with loss of production, supply, or human life, (4) fire, and (5) explosion or other emergencies leading to disaster affecting the public. The level of reporting also must be mentioned in the procedure. The report may include location, time, first witnessed by, details of incident, action taken, reporting authority, internal investigating authority, external investigation, and so on. The report also must be submitted to the regulatory authority. Lessons learned from the analysis of report are shared across the organization and the industry.

The networks are designed and operated to take care of any emergency in the city area as described. However, ensuring safety at all times calls for continuous monitoring, review, and improvement in the system. Monitoring, review, and improvement are the tools that not only improve the effectiveness and safety of existing system but also help in the design and implementation of more effective systems in future.

**ILLUSTRATION**

A study was undertaken to analyze the performance of three O&M service centers of a natural gas distribution company through service quality analysis. The studied company serves more than 25,000 customers having a PNG connection. Since the natural gas is an essential commodity for domestic, industrial, and commercial customers, the utility has to be efficient in their O&M services. To analyze the service quality, customer complaint registers maintained by the company were used as a secondary source of information. The complaints received during the specific year were used for analysis. The complaints were categorized based on its type and nature. The categories (type and nature) of complaints identified for three O&M centers were 24, 25, and 17. All the three centers received more or less similar type of complaints. Various causes of complaints (with numbers) for each category were also identified for all the three O&M centers using quality control tools (i.e., pareto chart, histogram, cause-effect diagram, checklist, and so on). Furthermore, the root cause analysis of different complaints was undertaken to facilitate efficient operations.

*(Continued)*

**ILLUSTRATION (Continued)**

Performance of service centers was analyzed using mainly three key performance indicators (KPIs) (i.e., average complaint response time, average complaint solving time, and average complaint disposal time). The three O&M service centers varied in their service performance. The service performance of O&M centers was then compared with the standards prescribed by the regulatory authority.

The complaints were further classified based on their timings of receipt (i.e., general shift, shift I, shift II, shift III) and compared with allocated manpower in different shifts. It helped in comparing the service levels of O&M centers with manpower resources available at each center.

**FINDINGS AND ANALYSIS**

1. The monthly complaint analysis revealed that the complaints are uniformly distributed across the month at all the three O&M centers.
2. It was found that maximum numbers of complaints were related to leakage of gas.
3. Complaint analysis by shifts revealed that about 77.5 percent of the total complaints are received during the general shift.
4. The average total time spent for one customer complaint was 39 minutes and 11 seconds, 55 minutes and 48 seconds, and 44 minutes and 31 seconds at three O&M centers, respectively. The average total time spent for one complaint for all the three O&M centers is 47 minutes and 7 seconds.
5. The average time required for solving the customer complaint after the receipt of complaint (i.e., average complaint disposal time) was 31 minutes and 59 seconds (based on 1,231 customer complaints received at all the three O&M centers).
6. It can be observed that the performance of the studied O&M service centers of the natural gas utility was excellent in comparison with the expected service level standards by the regulatory authority. The range of average complaint disposal time across all three O&M centers varies between 5 minutes to 3 hours and 5 minutes.

(*Continued*)

**ILLUSTRATION (Continued)**

**SUGGESTIONS**

Based on the complaints analysis the following suggestions may be considered by the utility.

1. Number of complaints received in general shift (i.e., between 9 am to 5 pm) at one of the O&M center is the least among all the three locations, whereas the manpower available is highest (10 persons compared to 6 in the other two locations). Feasibility of minimizing the manpower at this center may be investigated.
2. It is observed that the least number of complaints are received during the third shift. However, at two of the O&M centers, manpower allocated is same for all the shifts. Hence, the manpower allocation of technicians and helper and labour at these two centers for third shift must be studied for optimizing the operations.
3. It is observed that there was a significant difference in the average complaint solving time (time spent by O&M team in analyzing the complaint, finding the reason for complaint, and solving the complaint after reaching the customer site) among the three O&M service centers. Furthermore, complaint solving time for gas leakage which is the most common complaint across the O&M centers also varies. In both these parameters, efficiency of one of the O&M team was the highest and the other O&M team the lowest. The utility must develop a standard operating procedure (SOP) to solve complaints. An SOP improves the efficiency of technicians and also improves the complaint solving time. The utility can thus analyze the capabilities of work force for low efficiency and differences in the efficiency of the three O&M centers. If required, training needs can be identified and training can be imparted to improve the efficiency of the work force at the least efficient centers.
4. It was observed that at all the three O&M centers maximum numbers of complaints received were concerning "gas leakage." The proportion of false gas leakage complaints are 11.4, 50.4, and 28.2 percent of the total gas leakage complaints received at three O&M centers. This can be avoided if the company supplies odorized gas, which can help customers

*(Continued)*

**ILLUSTRATION (Continued)**

detect a gas leakage. The utility company can plan a campaign to educate customers about how to detect the gas leakage.

5. This analysis of customer complaints can be used to set up new O&M centers planned at future locations. It can help allocate manpower effectively and offer better services to customers.

   To optimize manpower and other resources such quality service analysis will be useful for natural gas distribution businesses.

(*Source*: Yadav, S., "Service quality analysis: Case study of natural gas utility," *Conference Proceedings (ISBN: 978-99937-51-38-0) 7th International Conference on Services Management* (ICSM -7), Institute of Tourism Studies, Macao SAR, China, 2014.)

## Bibliography

Gupta, A. (2009). "Preparedness to handle emergency in city gas distribution networks," *4th Pipeline Technology Conference.*

http://www.pngrb.gov.in (accessed on May 3, 2018).

Muhlbauer, W. K. (1996). *Pipeline Risk Management Manual*, Gulf Publishing Company, Houston, TX.

Paliwal, P. et al. (2010) "City gas distribution: Infrastructure and operations perspective," *Vikalpa-The Journal of decision Makers*, 35(4).

Yadav, S. (2014). "Service quality analysis: Case study of natural gas utility," *Conference Proceedings, 7th International Conference on Services Management (ICSM-7)*, Institute of Tourism Studies, Macao SAR, China.

Yadav, S., and Paliwal, P. (2011–2012). "Re-engineering service delivery process: Case of a natural gas utility," *Journal of Services Research (JSR)*, 11(2), 155–176.

# 7

# *Natural Gas Transmission and Distribution Business: Health, Safety, and Environment Issues*

## 7.1 Introduction

It is important to regulate health, safety, and environmental (HSE) issues for all natural gas pipeline systems—transmission as well as distribution. HSE management is critical in natural gas pipelines because it involves high-risk activities. Pipelines engage manpower, expensive complex machinery, and extensive support system that must be protected because these require high investment and have long payback periods. Cross-country pipelines are usually in remote areas and are prone to accidents, leakages, and damages. Similarly, though located in city areas, the natural gas distribution network is also vulnerable to various types of hazards and disruptions. Thus, if adequate safety measures are not considered, accidents can be sometimes catastrophic in terms of causalities, environmental pollution, loss of investment, and reputation.

## 7.2 Hazards Associated with Natural Gas Pipelines

Natural gas is a highly flammable substance, transported through cross-country pipelines at high pressure often close to dense population areas or through areas of high environmental sensitivity such as rivers, forests, lakes, and so on. The natural gas pipeline system poses severe hazard problems for populations and property close by. The provision of protective measures is essential for safe operation of a natural gas pipeline system. The requirement is due to:

1. The hazardous properties of natural gas
2. The quantity of natural gas, which could be released and its effect

**TABLE 7.1**

Causes of Pipeline Incidents

| Causes | Types |
|---|---|
| External interference | Digging, piling, ground works, anchor, bulldozer, excavator, sleeves, and so on |
| Corrosion | External, internal, and others |
| Construction defects | Construction or material related, welding defects, and component failures |
| Ground movement | Soil erosion, flood, landslide, mining, and so on |
| Others | Lightning and maintenance related |

In general, natural gas pipelines can be susceptible to damages such as holes, cracks, or ruptures.

Causes of incidents in pipelines can be broadly categorized as follows (Table 7.1).

Considering the fact that there are numerous possibilities for damages (and consequently the health and environmental after effects) being caused to natural gas pipelines, it is important to understand the framework that the natural gas industry deploys for management of HSE.

## 7.3 Tools and Techniques of Health, Safety, and Environment Management in the Natural Gas Transmission Pipeline System

HSE management involves identifying the potential hazards, assessing the risk from the individual scenarios, and evaluating the risk control measures to reduce the risk to humans, vegetation, and ecology. It also includes complying with the environmental regulations and minimizing pollution and waste to protect humans as well as animal life. Risk assessment is the relationship between probability and consequence (outcome) of an incident.

The generic methodology for risk mitigation is as follows:

- Collection of data for monitoring the pipeline network system
- Identification of hazards through systematic studies

- Relevant analyses for hazardous accident scenarios and subsequent modeling
- Appraisal of fire and safety management framework

Natural gas transmission companies as well as natural gas distribution companies implement a range of techniques—qualitative as well as quantitative—developed by technical, safety, and risk management professionals. The unidentified hazard may strike any time which may result in accidents and loss. So, it is very important that the hazard identification is carried out proactively in a comprehensive manner.

Although there are various techniques, two indicative techniques—hazard and operability (HAZOP) and computational pipeline monitoring (CPM)—provide an insight into the technological aspects of HSE management and are described in the following section.

## 7.4 Hazard and Operability Study

HAZOP (hazard and operability) study is one of the most commonly used hazard identification techniques that can be applied in various phases of a project management for a natural gas pipeline, including the front end engineering and design (FEED) phase, as a part of a detailed design phase and a plant operation phase or any modification or alteration of the plant. The HAZOP study is a semi-quantitative analysis and is used to identify deviations from the conceptualized design that could lead to hazards or operability problems and to define actions necessary to eliminate or mitigate these hazards and problems. The HAZOP study requires information such as process data, technical information, process and instrumentation diagrams, material balance sheets, process parameters, instrumentation diagram, site plans, line arrangement, list of safety valves, and so on.

## 7.5 Computational Pipeline Monitoring

Computational pipeline monitoring (CPM) techniques take information from the pipeline network related to pressures, flows, and temperatures to estimate the hydraulic behavior of natural gas. After the estimation is completed, the results are compared with other field references to detect the presence of an incongruity or unanticipated situation that may indicate a leak.

## 7.6 Engineering Tools Used for Safety in Pipeline Systems

Physical engineering tools like valves, coats, compressors, sensors and controls, pipeline inspection and repair technologies, leak detection mechanisms, and flow-rate quantification technologies are the most commonly used to increase the efficiency (and integrity) of the natural gas pipeline infrastructure for transmission and distribution.

A few tools that are used commonly are as follows:

**Pipe-coating materials:** Pipe coatings are applied to prevent the incident of corrosion underground as well as over the ground. In addition, as also mentioned in a previous chapter, cathodic protection often is used, which is a technique that involves the fundamentals of electricity to prevent corrosion.

**Valves:** Valves manage the physical flow of natural gas pipelines according to requirements and are the most important tooling system for handling emergency shut down and maintenance tasks.

**Leak detection techniques:** Various technologies for leak detection and methane leakage assessment that involve spectroscopy, laser spectrometry, external cable leak detection systems, drones for remote sensing detection, flame ionization detection, electrochemical detectors, and acoustic techniques are deployed for identifying leaks so that subsequent corrective actions can be carried out.

**Supervisory Control and Data Acquisition (SCADA):** These systems are the most sophisticated and well-accepted mechanisms for management of HSE, including pipeline integrity in natural gas pipeline systems. As also mentioned in one of the previous chapters, these systems are essentially sophisticated communications systems that measure and collect data along the pipeline (usually in metering or compressor stations and valves) and transmit the data to a control center. Data such as natural gas flow rate, operational status, pressure, and temperature are continuously collected for use to assess the status of the pipeline at any time. SCADA works in real time, so there is hardly any time gap between taking on-line measurements and transmitting them to the control center. This information helps concerned technical personnel to understand pipeline status on a continuous basis and enables swift actions in response to all possible malfunctions, thus leading to a robust monitoring and control system.

## 7.7 Organization Culture and Health, Safety, and Environment

Apart from providing for all sophisticated tool and techniques, equipment, following best practices in operations and maintenance (O&M), and adhering to tenets of asset integrity management for HSE management, it is equally important to have a company-wide strategy to make HSE a top priority around the clock. There must be an organization-wide strategic commitment to fostering a robust and positive safety culture. This commitment can be achieved by continually modeling and reinforcing a strong safety culture, along with effective processes and systems in which natural gas transmission and distribution companies strive towards the goal of zero accidents, hazards, and losses. Following two exhibits describing the commitments from two leading organizations—one from the natural gas transmission business and the other from the natural gas distribution business—clearly demonstrate this (Exhibits 7.1 and 7.2).

### EXHIBIT 7.1 HEALTH, SAFETY, AND ENVIRONMENT, AND SOCIAL IMPACT POLICY OF MOL GROUP, HUNGARY, ENGAGED IN A NATURAL GAS TRANSMISSION BUSINESS

HSE and social impact policy is gradually being implemented through a process of breaking down its goals into long-term objectives and local and group-level annual HSE objectives. Through this process we can ensure that all group units and members are operating in line with the principles described in our policies.

- Acting responsibly on the HSE and social impact of our activities as part of daily business
- Improving asset integrity and preventing incidents of every type while maintaining a high standard of emergency response
- Reducing our environmental footprint, protecting natural values, and supporting international efforts that address climate change-related risks
- Making a positive impact while eliminating negative impacts on the communities in which we operate, and on society in general
- Promoting a culture in which all MOL Group employees share these commitments

(*Source: https://molgroup.info/en/sustainability/sustainability-and-mol/hse-policy-and-strategy*)

### EXHIBIT 7.2 HEALTH, SAFETY, AND ENVIRONMENT POLICY OF MAHANAGAR GAS LIMITED (MGL), AN INDIAN CITY GAS DISTRIBUTION COMPANY

MGL is supplying piped and compressed natural gas to its stakeholders. Our Goal is zero injuries because we believe that injuries are preventable. We conduct our business in a responsible manner while adhering to internationally accepted good practice. HSE performance is everyone's responsibility and each one of us has a duty to intervene to prevent unsafe actions and to reinforce good behavior through demonstrating HSE leadership.

At MGL we are all committed to:

- Pursue the goal of no harm to people
- Protect the environment
- Use material and energy efficiently to provide our products and services
- Respect our neighbours and contribute to the societies in which we operate
- Develop energy resources, products and services consistent with these aims
- Publicly report on our performance
- Play a leading role in promoting best practice in our industries
- Manage HSE matters as any other critical business activity
- Promote a culture in which all MGL employees share this commitment

In this way we aim to have an HSE performance that we can be proud of and earn the confidence of customers, shareholders and society at large, apart from contributing to sustainable development including prevention of pollution.

In implementing this policy we:

- Have a systematic approach to HSE management designed to ensure compliance obligation with the law and MGL Life Saving Rules and to achieve continual performance improvement
- Set targets for improvement and measure, appraise and report performance

(*Continued*)

**EXHIBIT 7.2 (Continued) HEALTH, SAFETY, AND ENVIRONMENT POLICY OF MAHANAGAR GAS LIMITED (MGL), AN INDIAN CITY GAS DISTRIBUTION COMPANY**

- Require the contractors to manage HSE in line with this policy
- Engage effectively with neighbours and impacted communities
- Include HSE performance in the appraisal of staff and reward accordingly

NB: The Policy is made available to the public and interested parties.

(*Source: https://www.mahanagargas.com/health-and-security/mgl-hse-policy.aspx*)

The two exhibits reflect the commitment flowing right from the Chief Executive Officer (CEO) towards commitment to HSE management by following best practices and creating an organization-wide culture for HSE.

## 7.8 Conclusion

Addressing HSE concerns are crucial for natural gas transmission and distribution companies, not only for best standards of operations management and minimizing downtimes, but also to act as a responsible corporate citizen existing in harmony with all stakeholders.

## Bibliography

https://www.esri.com/en-us/industries/petroleum/segments/hse
http://www.hse.gov.uk/gas
https://www.inpex.co.jp/english/csr/pdf/sustainability2015-e11.pdf
www.naturalgas.org
www.pngrb.gov.in

# 8

# *Marketing Aspects of Natural Gas Transmission and Distribution Business*

## 8.1 Introduction

As discussed, to increase the natural gas share in overall energy mix, it is important to understand that natural gas transmission companies play a major role to reach end users of natural gas, which they facilitate through their vast cross-country pipeline networks. At times transmission and transportation entities also play key roles in natural gas sourcing through long-term, mid-term, and spot contracts with sellers in producing countries. Otherwise, most of the time their role is to operate as a transporter for the entities who own natural gas to supply it to desirable destinations over the transmission pipeline network.

Hence, after a discussion on regulations, applied technological issues, project management, operations and maintenance (O&M), and financing of natural gas transmission and distribution projects in the previous chapters, it is equally important to contextualize the downstream part of the natural gas value chain. This part of the value chain deals with marketing, pricing, customer service, and related issues. An attempt has been made to discuss these issues separately for transmission and distribution activities. That is because all marketing aspects of the natural gas transmission business mainly has a business to business (B2B) context whereas the natural gas distribution operates largely in a business to customer (B2C) framework.

## 8.2 Marketing Issues of Natural Gas Transportation Entities

Before delving into discussion on marketing and customer management aspects of natural gas transmission, it is important to have an idea about market segmentation relevant to the natural gas transmission business.

**Market segmentation natural gas transmission companies:** Natural gas transmission business is generally segmented by:

1. Type of industry
    - Power customers
    - Fertilizer customers
    - City gas distribution (CGD) customers including compressed natural gas (CNG)
    - Petrochemical plants and refineries
    - Other bulk industrial customers
2. Supply by source
    - Domestic gas (locally produced) customers
    - Imported natural gas through international pipelines/imported regasified liquid natural gas (R-LNG) customers

Each segment customer has a different need and hence transmission companies must consider these in a mutually value added manner. From the marketing point of view, transportation tariffs are a major mutual concern for natural gas transmission companies and its customers. In addition, the following aspects related to marketing are equally important between natural gas transmission companies and their customers:

- Managing activities related to customer service, contract renewal negotiations, and capacity marketing
- Consistent expansion of cross-country and international natural gas pipeline networks in uncovered regions
- Cultivate business relationships with existing and prospective customers
- Delivery of desirable qualities and quantities of natural gas at requisite pressures
- Taking care of uninterrupted deliveries
- Helping resolve technical and operational issues
- Designing mutually acceptable supply and purchase contracts considering market dynamics
- Create and develop new products and services through optimization of existing assets
- Tariffs and industry regulations information sharing to ensure corporate compliance while providing customers with an appropriate level of service

- Initiation and support of regulatory activities, including tariff changes, certificate filings, compliance filings, and so on
- Mutual engagement of marketing staff and/or customers towards developing and implementing both transportation and storage utilization (wherever applicable) strategies
- Addressing tariffs, billing, credits, and payment related issues

Pipeline laying activity considering future market expansion is one of the biggest challenges for natural gas transmission companies as they are in constant need for developing natural gas pipeline infrastructure. Land acquisition and right of way (RoW)[1] and right of use (RoU)[2] issues are critical for natural gas transmission companies.

Despite the B2B context of operations, all natural gas transmission entities attempt to provide the best customer assistance to their clients. With the use of modern monitoring technologies like SCADA, drones, GPS navigation, leak detection systems, and mobile apps, the transmission companies address all relevant customer service issues in the most desirable way. Moreover, all of them have interactive portals and well-qualified customer service professionals who also take care of these needs.

## 8.3 Marketing Issues for Natural Gas Distribution Business

As mentioned natural gas distribution operates in B2C environment and hence the marketing issues of LDC/CGD require a slightly different focus. LDC/CGD is also like any energy utility. The following section discusses the marketing issues of CGD business with the following structure:

- CGD market segmentation
- Build-up of the piped natural gas (PNG) selling price
- Market development challenges for CGD markets
- Fuel economics and CGD market development
- Customer service aspects
- Challenges for CGD business

[1] Right of Way (RoW) is a narrow piece of land granted, through an entry right or other means for laying natural gas pipeline.

[2] Right of Use (RoU) is a provisional right enjoyed by the proprietor of a natural gas pipeline network to use the piece of land obtained by RoW for laying pipelines.

### 8.3.1 City Gas Distribution Market Segmentation

As discussed, the CGD network is an interconnected network of gas pipelines. The associated equipment used for transporting natural gas from a high-pressure trunk line to the medium pressure distribution grid and subsequently to the service pipes supplying natural gas to domestic, industrial, or commercial premises and to the CNG stations situated in that specified geographical area (GA) comprises the total CGD outlay.

Basically, there are broad types of market segments in a CGD network: PNG and CNG. The former comprises domestic, commercial, and industrial customers whereas the latter is the vehicle fuel segment (Figure 8.1).

1. **PNG:** Provides natural gas to the domestic, industrial, and commercial segment.
   a. **Domestic PNG:** The segment of PNG which supplies gas to household customers. PNG has reduced the use of liquefied petroleum gas (LPG) and non-commercial fuel and is ushering in a healthy environment for cooking.
   b. **Commercial PNG:** Includes the supply of natural gas to hospitals, hotels, and restaurants and to industrial sectors which include small-scale business units.

   In the commercial sector, city gas is found to be very useful in applications like cooking, air conditioning, and power generation. The concept of combined heat and power (CH&P) is getting

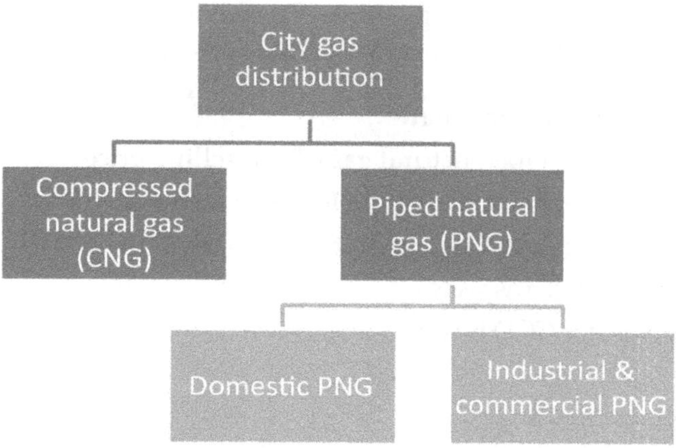


**FIGURE 8.1**
Segmentation for a CGD business.

popular because the system has much higher thermal efficiency as compared to other conventional services.

c. **Industrial PNG:** Includes power plant, fertilizer plant, and other industry. It is use as source of fuel for boiler and other heating equipment. It also used as feedstock in fertilizer plants.

2. **CNG:** Provides natural gas to the automobile segment by compressing natural gas through a compressor. There are CNG stations which dispense CNG. The use of natural gas in the transport sector has contributed significantly to the reduction of pollution. Apart from this, natural gas as CNG has replaced high-priced hydrocarbon commodities such as petrol and diesel, further contributing towards the cause of clean automotive fuels.

#### *8.3.1.1 Compressed Natural Gas Station Types*

Depending upon the structure and operations there are four major types of CNG stations:

1. **CNG mother stations:** They are connected to the pipeline and have a compressor of high capacity. They dispense CNG to an automobile through a dispenser. They also supply the CNG to a daughter station and CNG daughter booster station through mobile cascades. This type of station requires heavy investment.
2. **CNG daughter stations:** Mobile cascades from mother station come here. CNG from these cascades is dispensed through a dispenser to automobiles. After pressure in the dispenser falls, it is then replaced through a new cascade. It has the least investment compared to all types of CNG stations.
3. **CNG daughter-booster stations:** Drawbacks of daughter stations gave rise to daughter-booster stations. The cascade is a connected compressor, thus even though pressure in the cascade falls, it dispenses CNG at a constant pressure of 200 bar. Thus, it has the capacity of taking variable pressure and dispensing at a constant pressure. Developing a daughter-booster station is costlier than a daughter station as it has a compressor in it.
4. **CNG on-line stations:** CNG must be filled in an automobile at the high pressure of 200 bar. Thus, this station compresses the natural gas at a high pressure of around 250 bar and then dispenses it. Here the capacity of a compressor is less because it does not have a cascade facility to transport the gas to a daughter booster or a daughter station. Thus, overall investment is less compared to a CNG mother station (Figure 8.2).

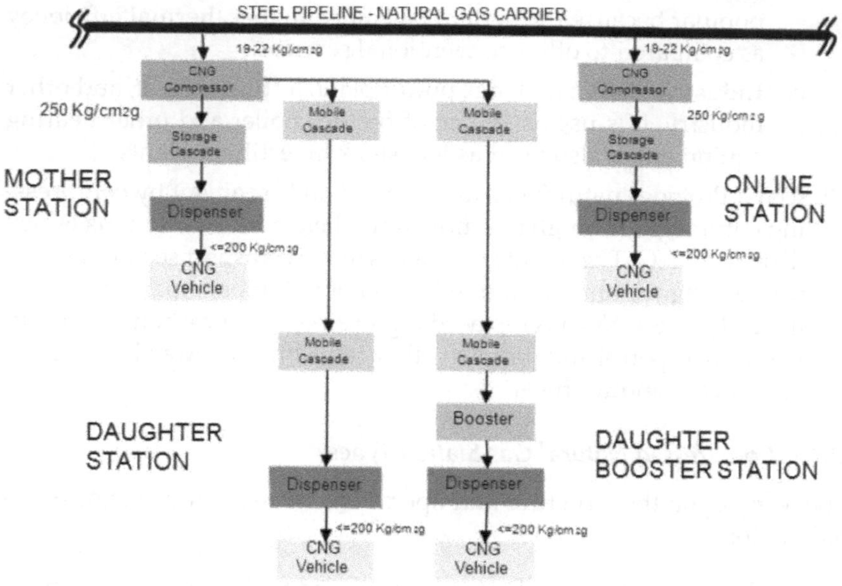


**FIGURE 8.2**
Types of CNG stations.

### 8.3.2 Build-Up of the Piped Natural Gas Selling Price

There are four major components that define the final selling price of PNG for the end consumer.

**Natural gas cost:** CGD companies buy natural gas from various sources. These sources may be local or based abroad. Depending upon the regions and markets, the imported natural gas is routed either through international pipelines (largely in Europe, Eurasia, Russian region) or comes as liquid natural gas (LNG). Prices of natural gas are normally expressed in U.S. dollars per million British thermal units ($/mmbtu). CGD companies may engage in various types of pricing arrangements with their suppliers ranging from term contracts to buying in spot markets. Depending upon their (CGD companies') PNG selling arrangements with consumers, fluctuations in natural gas sourcing price may reflect in PNG prices at the consumers' end.

**Network tariff:** Network tariff is the charge for utilizing the existing infrastructure of the authorized CGD operator. As discussed in the Chapter 2, network tariff being an important component in final PNG prices must be determined in such a way that it balances the interest of both entities: the network operator or marketer and the consumer.

**Compression charge:** Compression charge is levied on compression of natural gas and is applicable only in the CNG segment. The authorized entities are expected to recover the huge investments made in the infrastructural development primarily through levying network tariff and compression charges.

**Marketing margin:** It is a minor charge accrued to the marketer for undertaking marketing activities and providing service to consumers. It is generally a small fraction of the overall PNG selling price and is governed by market norms and/or prevailing regulations.

### 8.3.3 Market Development Challenges for City Gas Distribution Markets

The viability of a GA must be ascertained from its consumer mix. The industrial and large commercial segments offer the benefits of greater pricing flexibility, lower customer management efforts, and larger volumes. From the industrial and commercial customers' perspective, use of gas offers various benefits like cost savings, environment friendliness (gas being a cleaner fuel), better quality energy source (critical in continuous process-based industries), low maintenance costs, and storage and operational convenience, among others. Given that each individual consumer accounts for very small quantities and the task of serving a large consumer base has its own complexities, the domestic segment is considered the least profitable segment.

The industrial and large commercial customers act as **anchor (initial and base) customers** for CGD companies and provide large volumes in the initial years even as the PNG (domestic) and CNG segments require several years to build commercially viable volumes. It is observed that to ensure viability of a CGD project in most of the cities, leaving aside big cities, not less than 70 percent of the anchor (base) load should come from the industrial & commercial segments of the consumer mix. In these medium-small cities, the cost of infrastructure creation and development for CNG and PNG must be largely shouldered by the industrial demand.

The following matrix provides an idea about the phase-wise development of PNG and CNG markets for any CGD company (Table 8.1).

As illustrated, over short to medium term, the domestic segment has a mixed nature of growth; as a proportion of total demand, it grows initially and then the demand flattens. Commercial demand as a proportion remains almost the same over the years (though as in all cases, in absolute terms it grows with the growth in overall volumes in market). The share of the industrial segment gradually does down over the years, although in any year it still accounts for the largest market share (and in absolute volume terms also substantial). Rapid industrialization and natural gas demand coming from small and medium enterprises (SMEs) and distributed power solutions (gas–based micro CH&P and so on) generally boosts industrial demand.

**TABLE 8.1**

Illustrative Demand Profile in a Small and Medium CGD Network

| Segment/ Year | 3[a] | 4 | 5 | 6 | 7 | Driving Factors |
|---|---|---|---|---|---|---|
| Domestic | 7% | 10% | 11% | 9% | 8% | Vertical vs. horizontal spread, density/km$^2$, and willingness to pay upfront deposits |
| Commercial | 5% | 5% | 5% | 5% | 5% | Hotels/restaurants /power laundries/ schools, and so on |
| Industrial | 74% | 62% | 57% | 56% | 55% | As replacement of existing fuels |
| Transport | 14% | 22% | 27% | 29% | 32% | Conversion to clean fuels, CNG retrofitting costs, bus population, and traffic pattern |
| | 100% | 100% | 100% | 100% | 100% | |

[a] It takes an initial couple of years to develop a nascent market; however, for an already functional market this could be year 1.

It is interesting to note that the transportation segment (CNG) exhibits gradual growth over the years. This is because CNG is considered a clean fuel and is also economical as compared to other vehicular fuels such as petrol and diesel.

### 8.3.4 Fuel Economics and City Gas Distribution Market Development

CGD develops largely with the help of four key factors: gas supply, infrastructure, regulations, and economics or drivers. The industry has a natural advantage in building market through replacing existing fuels in domestic, small industrial, commercial, and transportation markets. However, this replacement will depend on the relative price of gas with respect to competing fuels.

Thus, fuel economics is a crucial issue in business development for CGD companies. Despite the virtues of natural gas as a clean fuel, many customers are price sensitive and hence they consider relative economic advantage or disadvantage in choice of PNG or CNG over conventional fuels.

Normally two fuels (conventional and natural gas) are compared based on their energy equivalence and respective price per quantum of calories (i.e., price/1000 kilocalories). Arriving at calorific value is important as the physical and chemical properties of competing fuels are different (for instance petrol and diesel are liquids whereas CNG is gaseous; similarly, as compared to natural gas, coal is solid).

In certain cases, investment in redesigning fuel handling systems and retrofitting of equipment is required to shift over from conventional fuels to natural gas and CNG. In most of the cases, dual fuel systems are in place. However, if natural gas is not made available at competitive prices on a consistent basis, then customers may go back to using conventional fuels. In that case, any investment in retrofitting and so on may become a sunk cost and may lead to consumer dissonance. Thus, PNG and CNG marketers must find ways to avoid such situations for best business development efforts. These ways include not only making available natural gas and related services at competitive prices but also risk sharing in cases of substantial investments in fuel handling systems and retrofitting.

Here it would be contextual to mention an interesting case study undertaken a few years back. This case study pertains to new market development for a natural gas distribution company. Aspects like switching from conventional fuel to natural gas, investments in retrofitting, natural gas price competitiveness, customer service, role of natural gas marketer as a solutions provider, engaging intensively with the customer to jointly find solutions, customer management, and so on are germane to this case study. The following exhibit gives a glimpse of the detailed case study.

## EXHIBIT: CUSTOMERS' VALUE EXPECTATIONS AND SUPPLIERS' VALUE PROPOSITIONS IN DEVELOPING NEW SERVICES AND RELATIONSHIPS: CASE STUDY FROM NATURAL GAS INDUSTRY[3]

### EXCERPTS FROM CASE STUDY

A successful interaction between customers' value expectations (CVE) and appropriate supplier value proposition (SVP) requires intensive supplier engagement with customers. The case study provides insights into how a natural gas marketer (GASCO), a supplier firm in a new market, engages intensively with its customers in the ceramic industry cluster (CERACO) to create synergies through long-lasting mutually beneficial value by co-creating new solutions for its customers. The case also illustrates how intensive discussions between the value co-creators help in mapping the entire process of identifying and implementing a new solution (i.e., natural gas-based fuel solution) replacing previous solutions to the same problem. The case study also considers the pragmatic application of the CVE-SVP interaction in the value co-creation chain, particularly highlighting its centrality to developing new services, relationships and markets, especially in context of a natural gas distribution company. The case also illustrates how CVE-SVP interaction proposition can also double up as a market entry strategy for the new suppliers.

(*Continued*)

---

[3] Singh, Ramendra & Paliwal, Pramod, "Customers' value expectations and suppliers' value propositions in developing new services and relationships: Case study from natural gas industry," *International Journal of Energy Sector Management*, 6 (2), 2012.

**EXHIBIT (Continued): CUSTOMERS' VALUE EXPECTATIONS AND SUPPLIERS' VALUE PROPOSITIONS IN DEVELOPING NEW SERVICES AND RELATIONSHIPS: CASE STUDY FROM NATURAL GAS INDUSTRY**

The significance of the case study stems from the unique context of the case study (i.e., the natural gas distribution business). Distance between the supplier and consumers and creation of a natural gas transmission and distribution infrastructure form major contours of this business. Moreover, natural gas once explored and produced from the wellhead cannot be stored easily. Furthermore, the creation of storage infrastructure entails significant capital expenditure and given the physical and chemical properties of natural gas, considerable space is taken up to store the gas. Therefore, ideally end-to-end supply tie-ups must be in place between suppliers and customers before the first cubic meter of gas is sold to the customer. In the absence of such tie-ups, the entire business model becomes unviable as return on investment in creating pipeline network for transportation and distribution until last mile becomes inadequate. It is a challenge for any business marketer to operate under such a business model unless it finds a match between CVE and SVP for the target market.

The case study illustrates how the environmental variables such as competitive intensity that affects profit margins necessitates an industry (CERACO) to review internal cost-structure and manufacturing processes, and design elements of CVE. GASCO finds a latent opportunity for matching CVE while planning a market entry using PNG as total fuel solution, replacing LPG as a previous solution. However, without engaging the prospective consumers as an intrinsic part of the interaction process, it was neither desirable nor possible for it to create value for its prospective customers by matching CVE with its own value propositions. The case study maps the process of this interaction that lead to the development of the new service and new relationships.

### 8.3.5 Customer Service

Customer service issues for CGD business are like that of any energy utility. These issues can be largely categorized as following:

- PNG service installation and starting
- Extension of domestic PNG services to new equipments
- Uninterrupted supply of PNG and CNG
- Outage management

- Expansion of CGD network to new areas
- Leak detection and management
- O&M assistance
- Billing, payments, and credit management
- Retrofitting needs for commercial, industrial, and CNG customers
- Assisting customers in PNG and CNG conservation
- Shifting of connection
- Temporary stopping of connection
- Termination of connection

All CGD companies are quite responsive to these issues. Moreover, many markets have robust customer service regulations which CGD entities must follow scrupulously. But with most of the markets becoming liberalized, CGD entities are quite proactive in their customer services. They deploy modern technology to address all relevant customer services management issues. Their interactive portals, 24 × 7 customer assistance, virtual assistants, mobile apps, digital metering and billing services, usage of infra-red technology-based leak detection equipment, mobile vans with necessary hazard management equipment, and so on are all aimed at providing customers with the highest level of service.

### 8.3.6 Challenges for City Gas Distribution Businesses

On the one hand, while there are huge opportunities, on the other hand, the natural gas distribution business is not without its share of challenges. Some of these challenges are:

- **Competitive pricing of PNG and CNG:** This helps in not only acquiring new customers but also retaining the existing ones.
- **Removal of some entry barriers:** To encourage competition, this is important for new players in distribution business. Necessary conditions must be created to attract new serious players in the business so that customers get the best prospects.
- **Capital intensive business:** Creating CGD networks entails sizable capital investments. The build-up of volumes may be sluggish and customer penetration of realistic levels may also take much time. In addition, long gestation periods coupled with low volume off-take can adversely impact the CGD business. Hence, it is important that players in the CGD business also have opportunities and facilitation to get a decent return on investment.
- **Facilitation from governments and regulatory bodies:** A CGD business like any utility business expects adequate facilitation from

concerned governments, municipal authorities, and regulatory bodies towards clearance of any stipulations and/or coordination issues with other utilities like telecommunications, sewage, water, roads, and so on.

- **Peculiarities of establishing CGD networks:** Peculiarities of urban management pose their own distinctive challenges for CGD businesses. Scalability, O&M, digging, hazard management, and other issues must be tackled in sync with issues of urban management.
- **Managing safety norms:** Handling an inflammable commodity presents its own challenges and hence CGD companies must be quite convincing on issues of HSE management.
- **External factors:** There are significant external factors such as prices of imported natural gas, emerging regulations and policies, and other extraneous factors that may have an impact not only on functioning of CGD businesses, but also may impact the profitability of the CDG entity.

As natural gas industry aims to garner a reasonable share in the energy baskets of various markets, transmission and distribution companies with their entire paraphernalia are aiding towards all such endeavors. Marketing is an important function that has a direct impact on the dealings of transmission and distribution companies with their customers. Therefore, understanding and taking all necessary steps towards these aspects give transmission and distribution companies a competitive edge in markets.

## Bibliography

Paliwal, P. (2010). "City gas India roundtable 2010: Initiatives and challenges," *Vikalpa-The Journal for Decision Makers*, 35(4), 61–91.

Paliwal, P. (2012). "Consumer behavior towards alternative energy products: A study," *International Journal of Consumer Studies*, 36(2), 238–243.

Singh, R., and P. Paliwal. (2012). "Interaction of customers' value appraisals and suppliers' value propositions in developing new services and relationships: A case study from the natural gas industry," *International Journal of Energy Sector Management*, 6(2).

Singh, R., P. Paliwal, and S. Sakariya. (2011). "Prabhar oil company and distribution challenges in the Indian Lubricants Industry," *Emerging Market Case Studies*, 1, 1–14.

Yadav, S., and P. Paliwal. (2011). "Re-engineering service delivery process: Case of a Natural Gas utility," *Journal of Services Research*, 11(2), 155–176.

# 9

# *Innovation in Natural Gas Distribution: The Case of LNG Express*

## 9.1 Background

As we have seen, natural gas distribution by a local distribution/city gas distribution (CGD) company aims at the last-mile connectivity. This last-mile connectivity needs the prerequisites of initially an elaborate cross-country natural gas transmission and later an intense CGD network. Both networks need substantial capital investments and a considerable amount of time to set up. Moreover, these networks, as seen, are fixed in nature, and thus an example of the economic concept of high level of "asset specificity" (i.e., an asset that can be used for a specific purpose only). In addition, the aspects of challenges and complexities in pipeline project management, hazard and safety management, operations and maintenance (O&M) over large distances, and so on are also pertinent to the natural gas transmission and distribution networks. These features make the business of investing in and later laying of transmission and distribution a proposition with unique risks and return trade-offs. On the other hand, all customers wanting to use natural gas must essentially await the availability of these networks (especially CGD) in their area.

On the one hand, stakeholders, particularly governments, have been advocating to increase the share of natural gas in the energy baskets (due to the virtues of this fuel). On the other hand, network-related reasons always pose a serious constraint in achieving this objective, at least at a faster pace.

While undoubtedly the natural gas industry, regulators, and governments have been working towards these objectives, the natural gas industry has simultaneously also considering innovative approaches to reach the consumers in faster, less complicated, and cheaper ways. This approach is especially more relevant for regions that must depend upon natural gas to fulfill their requirements. Thus, the question as to how to reach customers at the earliest looms large over the horizon.

An answer to this question may come from tackling the inflexibility in natural gas transportation and distribution. Is it possible to have a model that puts the last-mile customer first? And can that model overcome the essential need to have pipeline networks for making natural gas available to different segments (industrial, domestic, and commercial including transportation fuel)? The downstream value chain of liquefied natural gas (LNG) is largely thought to address these questions.

One of the modes by which natural gas travels (without using international or cross-country pipelines) over long distances is by converting natural gas into LNG (at temperatures of about minus 160 degree Celsius) and then transporting it through marine routes in specially built (cryogenic[1]) tankers.

Cryogenic tankers are built to handle LNG at extremely low temperatures (minus 160 degree Celsius). As also mentioned previously, natural gas when cooled at such low temperatures becomes liquid by reducing 1/600th in volume. Such compact volumes enable huge quantities of natural gas to travel long distances with relative ease (at the same time only the temperature must be handled technologically and not the pressure). At the upstream part of the LNG value chain, the liquefaction process indeed requires an elaborate technological and logistical process with substantial investments. Similarly, building and operating LNG tankers is also a complex and capital-intensive process. In addition, LNG tankers must be received at LNG terminals that have the infrastructure for storage, if needed, for regasification (the liquid form of natural gas, [i.e., LNG]), and for converting it (the regasified LNG [R-LNG]) back into transmission pipeline networks intended for both wholesale and retail customers. The latter, for their last-mile connectivity, need the flow of natural gas through an intensive distribution network.

The necessity of the upstream part of the LNG value chain will always remain. One reason also being that LNG is a modern and technologically sophisticated alternative to transportation of natural gas over long distances through conventional international pipeline options (moreover, it is not always possible to build international pipelines for long-distance natural gas transmission).

However, some innovations at the downstream part of the LNG value chain (i.e., post the receipt of LNG at the LNG-receiving terminals and before regasifying it again) can address the need of another necessity

[1] Cryogenics is the science dealing with behavior of materials at extremely low temperatures.

(i.e., the dependence upon secondary transmission pipelines [cross country] and distribution networks to reach natural gas customers).

The case study of LNG Express aptly demonstrates this.

## 9.2 LNG Express

The objective of this case study was to present an alternative business model for serving natural gas customers. With all its significance in place, conventional natural gas transmission and distribution businesses need to be aware of healthy competition posed by such innovative business models (as discussed herein). It would be interesting in the future to observe how the natural gas transmission and distribution industry responds to such innovations.

Structure of the case study:

- Company and activity brief
- Business model
- Products and services
- Implications for conventional natural gas transmission and distribution business model
- Conclusion

### 9.2.1 Company and Activity Brief

LNG Express India Pvt. Ltd is one of the companies in the Cryogas Industry Group and is engaged in the business of handling complete cryogenic storage, transfer, transport, and LNG regasification operations. This company is based in Vadodara, Gujarat-India.

Cryogas Industries was founded in 1996 in India. The group has diverse activities in the fields of cryogenic equipments and handling of cryogenic liquids and gases (including LNG). Different companies in the group carry out design, manufacture, supply, and turnkey installations of cryogenic equipments to handle cryogenic liquids and gases. The solutions apply across the spectrum of activities of several industries.

### 9.2.2 Business Model

LNG Express is in the business of supplying natural gas (by the LNG route) to off-pipeline consumers (i.e., those customers who do not have access to natural gas distribution pipelines). In many cases, the absence of distribution pipelines is a direct result of non-existence of transmission pipelines. LNG Express provides customized solutions of natural gas to the last-mile customers. Unlike the traditional LNG business model (where LNG is received at particular LNG-receiving terminals, is regasified with application of heating, and then pumped into transmission pipelines intended to various types of customers), LNG Express handles LNG in original form as received at the LNG terminal. Furthermore, it stores LNG at its site (and at customers' sites according to need and contractual agreements) and regasification happens only when it is required. LNG Express uses its expertise in handling cryogenic materials (such as LNG) in various contexts. Storage of LNG does not require dealing with different pressures (LNG is stored at atmospheric pressure). However, it does requires dealing with extremely low temperatures—a core competency of the company.

The company claims to have the latest modern LNG Hub and Spoke[2] solutions, which is conceptualized, designed, engineered, developed, manufactured, and commissioned by one of its group companies at a facility in Vadodara that enables it to reach the last-mile customers faster and safer. The LNG hub and spoke model and its importance in reaching the last-mile customer either through caskets or LNG trucks or even by mini and micro-bulk LNG Cryogenic Liquid Cylinders gives LNG Express a distinct edge.

LNG Express India Pvt. Ltd. provides LNG, R-LNG, liquefied to compressed natural gas (LCNG)(compressed natural gas (CNG) from LNG), auto LNG, micro-bulk LNG, and marine LNG by integrating cryogenic technologies coupled with automated cryogenic liquid and gas handling systems ensuring complete safety and reliability. Wherever needed, the company provides complete turnkey solutions to deliver natural gas to last-mile customers safely, efficiently, and economically. The company has relevant initial permissions required from the Indian natural gas regulatory authority, the Petroleum and Natural Gas Regulatory Board (PNGRB).

### 9.2.3 Products and Services

LNG Express offers a range of LNG distribution-related products and services catering to various needs of different segments.

- R-LNG for industrial customers
- BOOM model

[2] The Hub and Spoke distribution paradigm is a form of transport topology optimization in which traffic planners organize routes as a series of spokes that connect outlying points to a central hub. This model compares with point-to-point transit systems, in which each point has a direct route to every other point, and which modeled the principal method of transporting passengers and freight until the 1970s. (https://en.wikipedia.org/wiki/Spoke%E2%80%93hub_distribution_paradigm).

- LNG for marine fuel applications
- LNG for earth moving equipment
- R-LNG for temporary and peak shaving requirement using portable solutions
- Customized LNG storage and regasification for heavy industrial consumers
- R-LNG for municipal corporations and CGD companies
- LCNG stations
- LCNG caskets for remote locations
- LNG for trailers and buses
- LNG in micro bulk LNG cylinders

A brief discussion of all these products and services follows.

#### 9.2.3.1 Re-gasified Liquid Natural Gas for Industrial Customers

The company supplies LNG regasification station or terminal on an outright sale basis for captive consumption to replace almost any fuel within the premises. For such an LNG regasification station, LNG Express provides express transport service to deliver LNG in required volumes to ensure uninterrupted consumption of natural gas. An automated programmable logic controller (PLC)-based LNG regasification plant is designed considering daily consumption, distance from the terminal, and application criticality, as well as maintenance of LNG back up stock.

In some of the geographical locations, LNG Express may have a mother station and can offer a smaller regasification system to ensure seamless and uninterrupted consumption, which eventually reduces the overall land required from the end user. LNG Express, thereby, enables its customers to achieve major objectives such as efficiency, cost competitiveness, safety, and reliability at the lowest life cycle cost (Figure 9.1).

#### 9.2.3.2 Build, Own, Operate, and Maintain Model

For customers who do not wish to invest in LNG equipment, the company offers complete an LNG regasification station on a build, own, operate, and maintain basis called BOOM.

#### 9.2.3.3 Liquid Natural Gas for Marine Fuel Applications

LNG Express provides an LNG fuel tank in fully compatible, qualified, and internationally certified systems to enable use of LNG for tug boats, short sea vessels, riverine transport vessels, large bunker barges, and articulated vehicles, as well as cargo vessels. LNG Express provides fully integrated LNG fuel stations for marine applications.

**FIGURE 9.1**
LNG storage tank at LNG express site. (Courtesy of IWI Cryogenic Vaporization Systems (INDIA) Pvt. Ltd., Vadodara, India.)

#### *9.2.3.4 Liquid Natural Gas for Earth Moving Equipments*

LNG Express in cooperation with its sister company Cryogas Equipment provides LNG for earth moving and mining applications. This arrangement is aimed at significantly reducing the consumption of diesel with maintaining low maintenance and low costs (Figure 9.2).

#### *9.2.3.5 Regasified Liquid Natural Gas for Temporary and Peak Shaving Requirement Using Portable Solutions*

LNG Express provides regasified liquefied natural gas (R-LNG) using fully integrated plug and play solutions for use of R-LNG at remote locations and one-time applications for satisfying peak demands. The mobile application system is also capable of supplying natural gas conveniently to islands and for barge-based bulk supplies (Figure 9.3).

#### *9.2.3.6 Customized Liquid Natural Gas Storage and Regasification for Heavy Industrial Consumers*

LNG Express provides fully dedicated captive consumption LNG regasification stations integrated with LNG transportation from nearby terminals

**FIGURE 9.2**
LNG fuel tank for earthmovers. (Courtesy of Cryogas Equipment Private Limited—Shop Picture, Vadodara, India.)

using LNG trailers, containers, or barge-based multiple and large volume transportation of LNG. This fully purpose-built solution is provided for islands, mega-consumers for power production, and other industrial applications as well as R-LNG for municipal corporations to prepare their markets in advance of reaching the pipeline.

**FIGURE 9.3**
Portable LNG solutions. (Courtesy of Cryonorm-BV Holland, Alphen aan den Rijn, the Netherlands.)

### *9.2.3.7 Regasified Liquid Natural Gas for Municipal Corporations and City Gas Distribution Companies*

LNG Express provides LNG for fully dedicated LNG storage and regasification modules for municipal corporations and CGD companies to provide the maximum utilization of pipelines and to support existing customers, and to increase gas consumption volumes. For such regasification stations, suitable dozing of mercaptane also is provided to meet the regulatory requirement.

LNG Express operates mother stations that are versatile LNG-based stations specifically developed for the storage of LNG and processing of stored LNG to cater to many applications, to replace different fuels such as petrol, diesel, liquefied petroleum gas (LPG), Naphtha, furnace oil , light diesel oil, and so on.

### *9.2.3.8 Stations Making Compressed Natural Gas from Liquid Natural Gas*

LCNG stands for making compressed natural gas (CNG) from LNG. LNG is processed at an LNG express mother station (LEMS) (main source station) using an LNG booster pump to make LCNG by boosting the LNG pressure from 2 to 6 bars (g) to 255 bars (g).

LEMS offers dispensing facility for auto rickshaws, cars, and buses or heavy vehicles. LEMS also provides filling of LCNG caskets, which enables dispensing of premium CNG (P-CNG in remote locations using the LCNG caskets with or without booster pumps. LCNG dispensed to remote locations through LCNG caskets is termed as P-CNG by the company.

An LNG mother station also has provision for refilling of micro bulk LNG cylinders using the latest technology for filling. LEMS also has provision for LNG dispensing for buses, trucks, and trailers that are installed with an LNG fuel tank either for single fuel or dual fuel. LNG is dispensed using on the fly (while in motion) system equipped with a fully automated LNG dispensing.

LNG express liquid-based daughter stations (LELDS; fed from other mother stations) are equipped with smaller capacity storage tanks between 3,000 and 28,000 liters. This smaller capacity facilitates serving customers in remote locations as well as in densely populated regions with equal ease without the customers having to have storage spaces.

The major advantages of LNG-based daughter stations are lower power requirement, lower or negligible maintenance, no need to use oil, and reducing the large waitlist at the mother stations. LELDS are quite flexible in nature and extremely cost effective (Figure 9.4).

As compared to CNG (which is compressed from the pipeline, and, by default, carries sulphur, moisture, carbon dioxide, and other impurities, and traces of oil from the compression), LNG provides the highest quality LCNG. This quality is possible because before liquefaction of natural gas, the sulphur, higher hydrocarbons, moisture, carbon dioxide, and so on have already been removed to facilitate the liquefaction process.

**FIGURE 9.4**
LCNG infrastructure. (Courtesy of LNG Express India Private Limited—LCNG station, Vadodara, India.)

#### *9.2.3.9 LCNG Caskets for Remote Locations*

There will always be remote locations where the availability of power and bar load may be an issue. LNG Express provides for LCNG casket-based daughter simple and daughter booster stations such that remote markets with smaller consumption are provided with P-CNG. And so, the last-mile customers in rural areas do not have to travel long distances to get the P-CNG (Figures 9.5 and 9.6).

**FIGURE 9.5**
LCNG caskets. (Courtesy of LNG Express India Private Limited—LCNG station, Vadodara, India.)

**FIGURE 9.6**
LCNG casket carrier. (Courtesy of LNG Express India Private Limited—LCNG station, Vadodara, India.)

### *9.2.3.10 Liquid Natural Gas for Trailers and Buses*

LNG Express provides LNG to specially developed LNG fuel tanks installed in one or multiple modules on trucks and buses. This LNG-fueling solution is available for light commercial vehicles and high commercial vehicles (Figure 9.7).

### *9.2.3.11 Liquid Natural Gas in Micro Bulk Liquid Natural Gas Cylinders*

LNG Express offers multiple types of micro-bulk LNG liquid cylinders to meet the requirements of different categories of customers such as restaurants, hotels, hospitals, institutions, canteens, shopping malls, and so on. LNG micro-bulk cylinders are available in 50, 200, 450, 750, and 1,000 liters. Most of these LNG micro-bulk cylinders have built-in pressure building system as well as regasification system to provide R-LNG to its users in a most user-friendly manner. Handling of LNG micro-bulk cylinders is one of the specializations of LNG Express where multiple cylinders are handled with specifically designed handling equipment and sophisticated logistical support. Incidentally, electric heating is not required for LNG regasification and hence energy is saved. LNG Express provides several sizes of such liquid cylinders, which are a replacement for LPG cylinders using the "fill to empty"

**FIGURE 9.7**
Heavy vehicles run directly on LNG. (Courtesy of Cryonorm-BV Holland, Alphen aan den Rijn, the Netherlands.)

concept. Major advantages of using LNG over LPG are higher fuel efficiency, lower replacement cost, negligible maintenance, full utilization, economy, and safe storage (Figure 9.8).

### 9.2.3.12 Liquid Natural Gas Dispensing Equipment for Mining Industry

Cryogas Equipment, a company within Cryogas Industries Group, is engaged in manufacturing and delivering turnkey LNG dispensing stations

**FIGURE 9.8**
LNG micro-bulk cylinders. (Courtesy of Cryogas Equipment Private Limited—Shop Picture, Vadodara, India.)

from its facilities in Vadodara, Gujarat, India, to operate mining dump trucks for its mining industry clients located in faraway places where using diesel for fuelling may be cumbersome and relatively expensive (Figure 9.9).

The dispensing unit consists of the following:

1. LNG saturation skid: Sub-cooled LNG is saturated before it is filled into the fuel tank (installed on board the mining truck). This procedure is to ensure safe operation of mining trucks in rugged conditions.
2. LNG dispensing pump skid: Ensures timely and quick refilling of on-board (of heavy trucks) LNG fuel tanks without time lags.
3. LNG dispenser: It is designed for adequate safety, flow measurement, and has a soft touch control panel.

Vacuum-insulated pipelines are used to transfer LNG from tankers to a storage tank and from a storage tank from the LNG dispensing pump to the LNG dispenser for fueling LNG-fueled heavy trucks. Cryogas uses a totally integrated PLC supervisory control and data acquisition (SCADA) system for

**FIGURE 9.9**
LNG dispensing unit. (Courtesy of Cryogas Equipment Private Limited & LNG Express, Vadodara, India.)

**FIGURE 9.10**
Heavy truck at mining site running on LNG. (Courtesy of Cryogas Equipment Private Limited & LNG Express, Vadodara, India.)

monitoring the process. Adequate care is taken to pre-test the unit using liquid nitrogen in the Cryogas Workshop before dispatching it to the customers to ensure safety and reliability of the system (Figure 9.10).

## 9.3 Implications for the Conventional Natural Gas Transmission and Distribution Business Model

As discussed in the previous section, LNG Express provides natural gas solutions that are also offered as piped natural gas (PNG)-based solutions by CGD companies. Moreover, the flexibility and versatility with which it offers its products and services have significant implications. On one hand, natural gas transmission and distribution companies can treat this as a threat to their own business model; on the other hand, they may see at least two areas where the business model of LNG Express (and similar companies) can also be complementary to their two important businesses (i.e., CNG and PNG). Instead of being threat, such innovative business models can act as additional supply sources to distribution companies. This outlook, in our opinion, increases the supply security for CGD companies.

## 9.4 Conclusion

Natural gas transmission and distribution companies must consider innovative business models in the natural gas supply business. These innovations will enable them not only to reinvent their own business models but also to cooperate with competing business models for mutual advantages and better customer service.

## Acknowledgments

The authors are grateful to Mr. Nayan Pandya, Director of Cryogas Industry Group and LNG Express Pvt. Limited, Vadodara, Gujarat-India, for permission, cooperation, and invaluable inputs towards writing this case study.

The authors also acknowledge the support from Mr. Varun Patel towards organizing LNG Express plant and facilities visit which went a long way towards the analysis of this innovative natural gas distribution model. All pictures are courtesy Cryogas Industries Group and LNG Express Pvt. Ltd.

***Disclaimer:*** *All information including figures, technological details and pictures in this case study is based on the inputs available in public domain, discussions with senior management team of LNG Express and authors' field visit to the company's works and office. Documentation of the case study has due approval of the management of LNG Express Pvt. Ltd. and Cryogas Industries Group. The case study has been written with an objective to stimulate academic discussion on emerging Natural Gas distribution models and authors do not attempt to certify the technologies, pictures, designs, products, and services mentioned in the case study.*

# Index

**Note:** Page numbers in italic and bold refer to figures and tables respectively.

Milton Keynes UK
Ingram Content Group UK Ltd.
UKHW040058071024
449327UK00019B/635

9 780367 656584